T0092273

IT Change Management
A Practitioner's Guide

London: TSO

a Williams Lea company

Published by TSO (The Stationery Office), part of Williams Lea, and available from:

Online
www.tsoshop.co.uk

Mail, Telephone, Fax & E-mail
TSO
PO Box 29, Norwich, NR3 1GN
Telephone orders/General enquiries: 0333 202 5070
Fax orders: 0333 202 5080
E-mail: customer.services@tso.co.uk
Textphone 0333 202 5077

TSO@Blackwell and other Accredited Agents

International Best Practice (IBP) is a framework-neutral, independent imprint of TSO. We source, create and publish guidance which improves business processes and efficiency. We deliver an ever-expanding range of best-practice guidance and frameworks to a global audience.

A CIP catalogue record for this book is available from the British Library

A Library of Congress CIP catalogue record has been applied for

First published 2017

ISBN 9780117083653

Printed in the United Kingdom for The Stationery Office

J003951839

Contents

List of figures

List of tables

Dedicated to my father, who taught me the pursuit of excellence

About this guide

Who's it for?

The short answer is – all change management practitioners. Whether you're new to this field or have been involved with it for years, you'll find this publication helpful in establishing, maturing and optimizing change management in your organization.

What you'll find here aren't random or theoretical thoughts about change management; the content is derived from decades of real-world lessons learned (often the hard way).

If you are a novice, this publication will guide you in how to introduce a very basic change management capability that will help your organization get better control of IT changes happening in your environment. I've addressed the full spectrum of issues you'll face in getting change management established and realizing business value.

For the experienced practitioner whose organization has an existing change management programme and is looking for help in maturing and optimizing, you'll find expert guidance here on how to do it while minimizing cultural resistance.

Why do you need it?

Change management, in theory, is relatively simple. In practice, however, it can be very challenging. This publication is based on the author's in-depth experience in change management, where things are often not as simple as the best-practice training might lead you to believe.

Who's it by?

Greg Sanker

Greg has decades of real-world IT experience, ranging from a global technology giant to a small state government agency. He has been involved with IT change management in various capacities over the years.

Greg is an IT service management practitioner and author who frequently speaks about organizational excellence and change management. He is noted for his practical approach and 'been there, done that' advice.

How is it organized?

This publication is structured to be useful in a wide variety of situations. If your organization is just getting started with formal change management, you'll benefit from a front-to-back reading. If your organization has basic change management, you'll need to pay particular attention to the basic programme outlined in Chapter 3 (phase 1). You may find that you're already meeting all the described criteria, in which case you're ready for the next phase (covered in Chapter 4). Chapter 5 on optimizing change management can be added as a third phase, or included as parts of earlier phases as needed, to meet the challenges you face in your situation.

A summary of the chapters is as follows:

- **Chapter 1** Introduces change management essentials and provides a basic understanding of what is meant by 'change management', what change management seeks to do for the organization and why it is important for organizational success.

- **Chapter 2** Addresses some of the challenges change management presents and describes how various approaches have contributed to change management failures and organizational frustration. It proposes a multiphased approach to adopting and maturing change management, which is then taken up in more detail in the following chapters.

- **Chapter 3** Provides guidance to allow you to identify and put the right controls in place to manage stability and uptime, and be proactive in minimizing any adverse business impact. This chapter focuses on introducing the concept of change control into an organization, with the emphasis on business value and addressing the cultural considerations necessary for success. It then sets the stage for future maturity.

- **Chapter 4** Establishes the idea of proactively managing the end-to-end change lifecycle by introducing a review stage at the end of development/testing to ensure changes introduced both support the objectives of the business and minimize the risk of an unsuccessful implementation. It then introduces the concept of optimization.

- **Chapter 5** Provides strategies for optimizing any change management programme, including standard changes, delegated change authority and change models. The practical guidance here is applicable to any change programme that is struggling to keep pace with the business.

- **Chapter 6** The successful change programme must fit the organization in which it operates. This chapter explores the foundational concepts of adapting and adopting as they apply to change management.

- **Chapter 7** Takes a brief look at what lies ahead for change management. DevOps and related 'iterative' and 'continuous flow' development methodologies challenge the traditional approach to change management.

Foreword

We hear criticisms of IT change management from many angles: it's too bureaucratic, too process-focused, too IT-focused and certainly not responsive enough to survive in a modern business environment in which IT plays an increasingly vital role. How, critics ask, can a change advisory board (CAB) review and evaluate every change, when businesses – increasingly dependent on IT – need to be moving ahead at lightning speed?

Change management, however, is much more than just the installation of a CAB; it is the planned construction of capability that informs and accelerates the maintenance and improvement of IT. Whether you are just beginning IT change management in your organization or are under pressure to make it deliver better outcomes more in sync with your multivendor cloud environment or new DevOps initiative, the insights and recommendations in this publication can help you accomplish your goals.

Yes, in today's businesses, successful changes need to happen more frequently and rapidly than ever, but that presents a challenge to change management, not a negation of it. Too often, service managers and developers think of change management as something that happens once – just before release and deployment. But it is not a last-minute quality check. It is change control as an integral part of every aspect of IT's mission, ensuring compliance and stability, and keeping changes focused on business needs and outcomes.

Change management, approached this way, aligns with multiple frameworks and methodologies. In fact, as Greg points out, IT change management must be done in the context of your organization's culture, regardless of the frameworks you have chosen.

Stability isn't the absence of change …

The definitions of standard, normal and emergency changes can reduce the burden on (and accelerate the function of) the CAB, which ceases to be a barrier and becomes an advisory partner. Maintaining a laser-like focus on quality, stability and compliance requirements throughout the development and/or acquisition of new or changing technology can minimize risk and reduce the frequency and impact of interruptions resulting from changes.

'Out of the box' is not the place to start, especially since technology and tools are secondary to a change management capability that suits the needs and priorities established by your business. Neither is 'by the book' the way to proceed, although Greg states that following best practices doesn't mean you can only achieve the average, or that you are copying other organizations.

Read what Greg's book has to say. Make this your guide to a better IT change management capability during this time of technological disruption and rapid business transformation.

Roy Atkinson
Senior writer/analyst, UBM Americas – HDI

Preface

The most commonly asked question when IT systems fail is, 'What changed?' We instinctively know something must have; otherwise, it would have kept on working.

Change-related failures pose significant risk to business. In response, organizations apply best-practice frameworks such as ITIL® and standards such as ISO 20000. The goal for these efforts: to gain better control of the IT infrastructure and improve operational effectiveness. Unfortunately, despite significant improvements, organizations continue to struggle to effectively manage IT changes.

Worse, much of what has been done under the umbrella of 'change management' is overly complex, too internally focused on IT and doesn't keep pace with business needs.

Some words I frequently hear in connection with a typical change management implementation are:

- Slow
- Bureaucratic
- Complex
- IT-focused
- Broken.

Here I present a practitioner's guide to establishing, maturing and optimizing an effective IT change management capability (or programme). It's intended to answer the question, 'How do I actually do it?' The goal is to help you be successful in your organization.

In contrast to a process-focused approach, my intention is to address the full spectrum of elements necessary for successful implementation, which include:

- Focus on outcomes (not process)
- Realize business value
- Adapt and adopt
- Gain (and maintain) organizational support
- Consider the culture (organizational change management)
- Phase in implementation with incremental improvements.

It's my hope that you will find this publication helpful as a daily guide to managing the challenges the change management practitioner faces.

Acknowledgements

I am deeply indebted to so many people for helping shape this publication – the countless change management practitioners the world over whom I've spoken to in person, at conferences or by email. Each one represents a unique set of challenges and circumstances.

I am grateful for the wealth of knowledge generously shared by Stuart Rance, Kirstie Megowan and Ivor Macfarlane. I am especially thankful for the extensive review and coaching of David Moskowitz, without whom this publication would not have happened.

Lastly, I would be remiss to not mention my long-time friend and mentor, Mark Thomas; without his advice and encouragement, my career in IT service management would not have been possible.

About the reviewers

Roy Atkinson

Roy Atkinson is HDI's senior writer/analyst, acting as in-house subject matter expert and chief writer for blogs, articles and white papers; he is also a member of the HDI International Certification Standards Committee. He co-authored the AXELOS/HDI white paper, 'Synergies between ITIL® and Knowledge-Centered Support'. He has a background in service desk, desktop support and small-business consulting. He studied advanced management strategy at Tulane University's Freeman Graduate School of Business, and holds HDI certificates in support center management (HDI-SCM) and knowledge-centered support (KCS).

Andie Kis

An ITIL, PRINCE2® and Green IT qualified senior consultant specializing in service integration and management, Andie works for Information Services Group (ISG) Ltd. She has a strong practical background in business relationship management, service desks and the people side of IT, having started her IT career in 2002 on a multilingual IT help desk in Budapest, Hungary. In Andie's own words, she is a practitioner of best practices with a chameleon-like ability to adapt to and understand a wide variety of customers and clients. She has worked in different industries in internal and outsourced roles, working with the likes of TCS, EDS, the BBC, Deutsche Bank and Macmillan Cancer Support. She is an active international speaker as well as a contributor to various industry magazines and blogs.

David Moskowitz

David Moskowitz is an IT service management consultant with more than 30 years' experience of working with clients to help them improve outcomes that engage both business and IT. He has written, edited and taught ITIL, PRINCE2 and RESILIA™ courseware and articles. He is a systems thinker and certified as an ITIL Expert.

Peter Saxon

With more than 15 years' professional experience in an IT environment, covering areas such as retail, logistics, NHS and automotive, Peter Saxon has held various roles, including those of computer analyst, service desk manager, operations manager, project and programme manager, and business change manager. Each of these roles has involved a significant amount of IT and business change management.

Introduction

Change is everywhere. Everything is in a constant state of flux.

Businesses used to have one choice when it came to technology – their internal IT staff. But with mobile computing and cloud services infrastructure on demand, we live in a new era where anyone anywhere in the world with a bright idea can make it available to the global marketplace in a few hours. This 'consumerization' of IT services puts enterprise-class IT services at the fingertips of everyone in your company.

All modern enterprises depend on the digital infrastructure for their very lifeblood. The game is no longer simple decision support; entire businesses depend on unprecedented strategic advantages afforded to those who master big data and can move with agility to capitalize on the narrow window of opportunity earned by their advantage.

Likewise, both government and public IT sectors face similar pressures to adapt. Many agencies are still operating with a 1950s mindset, automated by technology in the late 1970s. Meanwhile, a generation of 'digital natives' expect public agencies to provide services and access to information in the same ways they've come to expect from private businesses. They don't understand why it's so difficult, and have no interest in the complexities of legacy IT systems. As organizations struggle to meet demands for modern services with antiquated systems, the news is full of IT project failures and alarming security breaches.

IT systems represent an increasingly complex hybrid of multivendor applications, services and infrastructure. IT organizations face cloud and other forms of 'as a service' delivery models that simply didn't exist when even recently hired staff were in college. The world has never been more connected, enabling instantaneous global collaboration with unprecedented levels of complexity, operational challenges and risks. Hardly a day passes without the discovery of a new security breach, a vulnerability exposed or personal information compromised.

IT organizations are facing increased regulatory and compliance pressures as governments and industry organizations address emerging risks and challenges. The complexity and rate of change can be overwhelming, and there's a temptation to want things to stay the way they are because we understand the current state and are comfortable with it. But such 'stability' is an illusion. Stability isn't the absence of change but rather the deliberate application of the right changes that restore balance and maintain stability.

Wise is the IT organization with a strong and effective change management capability that meets the needs of its customers.

> **Example**
>
> A US government agency commissioned a major application development effort. The contractor gathered requirements and managed the project through to completion. In line with the agency's usual practice, when the development was complete, the change was brought to the change advisory board (CAB) for sign-off before being released into production.
>
> Much to the surprise of the project team, the change was denied. Highly offended, they demanded to know what right the change manager had to block such an important change. Regretfully, the change manager had to tell the project team they had developed the new application for a platform that the agency didn't have in the production environment.

All change involves risk: risk of breaking something, losing something or not achieving the desired results. IT change management is the art and science of effective and efficient implementation of beneficial changes to IT services at the speed business requires, while minimizing negative impact and managing risk.

Many IT organizations have adopted some form of IT service management (ITSM) best-practice framework, at least to a degree. ITSM training continues to be very popular, with numerous qualification and certification schemas. These are excellent sources of information, and I highly recommend this type of training. However, if it's a change management process you're after, you can have that in a week. On the other hand, if you want an *effective* change management capability, well, that will take a bit more time.

An effective change management capability requires more than just training in best practices and process frameworks, because it is far more than merely implementing a process for managing change. It requires a great deal of knowledge and understanding of the particular organization and its unique challenges. The change management practitioner must understand the organization's culture, challenges and goals.

And it goes even deeper than that. Change management cannot be just the same thing only faster; it must be engineered from the ground up to meet the current and ever-changing needs of business. Best practices, including change management, must be adopted *and* adapted.

The challenges facing the change management practitioner have never been greater, yet there's never been a better time to be a change manager. Successful organizations require highly effective IT change management. So where do we begin?

We start right here – with the end in mind. In this publication you'll find a very straightforward multiphased approach to make change management work in your organization. It is structured in logical steps, or phases, that focus on the success of each individual step while maximizing the likelihood of building a mature change management capability.

What outcomes do our customers need that IT change management can help facilitate? How will our change management capability enable the business to be more successful? These are the kinds of question that should be at the forefront of your mind as a change management practitioner, and this publication willl help you work through the answers.

1 The case for change management

> *Change is the only constant in life.* Heraclitus

Change management is arguably the most widely adopted IT service management (ITSM) practice. Whether it is formally recognized or not, every organization has some form of change management because it is simply impossible to maintain IT services without managing changes. Unfortunately, there are nearly as many definitions of change management as there are those who discuss it.

1.1 Change management goals

Let's start with a very basic definition of change management. What is it tasked with accomplishing?

IT change management is an organizational capability that seeks to:

● Support timely and effective implementation of business-required changes

● Manage risk to the business appropriately

● Minimize negative impact of changes to/for the business

● Ensure changes achieve desired business outcomes

● Ensure governance and compliance expectations are met.

You'll quickly notice that these goals do not sound very IT-like, at least in the traditional sense. They have little to do with technology and infrastructure and much to do with what the business requires.

Managing the technical details of IT changes is necessary and important to ensure successful business outcomes. Unfortunately, a change management programme that starts with an internal (IT) view has limited scope and is the root cause of many such programmes struggling or failing altogether.

Change management is often referred to as a process (and there are most certainly processes involved with managing changes), but to be successful it needs to be much more than that. Successful change management is an organizational capability that supports the changing business needs, and to that end, it must start and finish with a singular focus on achieving business outcomes.

Let's take a closer look at the goals of change management.

1.1.1 Timely and effective implementation of beneficial changes

This is a pretty straightforward goal, but from it, two key principles emerge.

1.1.1.1 Timeliness

When it comes to timeliness, the IT organization's objective is to support the business by harmonizing with (not outpacing) the rate of change needed by the business to support its goals.

In highly competitive markets, business is driven by a strong sense of urgency: the need to get products quickly to market to maximize profit, and to innovate faster than the competition to gain market share. These companies' IT organizations must match the rate of change required to enable that competitive advantage.

But faster is not always what's required. For instance, if your organization builds and maintains aircraft or bridges or develops products for healthcare or the military, timeliness takes on a different meaning. In these cases, change management must match the need for precision and flawless execution. Product development cycles in these industries and the nature of the businesses create different demands on IT's change management.

It's not a matter of right and wrong – it's about understanding the business you serve, and building and adapting capabilities to the meet its needs.

1.1.1.2 Effectiveness

Effectiveness is a measure of how well the change management capability meets business needs. The end result of effective change management is changes that are consistently implemented when needed and that produce the expected business outcomes.

Effective change management must be mindful of 'time to value' – the time from identification of the need for a change to the time the business value from the change is realized. Notice that the end point is not when the change itself is implemented, but rather when the change produces the desired (or expected) business outcome. Some changes never achieve the desired result, which makes time to value infinite – the opposite of what's desired.

This end-to-end view of change effectiveness is what differentiates a change management capability from a change management process. It is also a departure from the typical technology and IT-centric view taken by many change management programmes.

1.1.2 Manage risk

It probably goes without saying that IT change management comes with an implicit oath to 'do no harm' to the business (which includes the underpinning IT services). But it is also true that all changes come with some degree of risk. Successful change management must balance these two aspects such that the business enjoys the outcomes the change promises with no, or minimal, harm.

When things go wrong in IT, the most frequently asked question is, 'What changed?' The capability to understand and manage change-associated risk is a large part of effective change management. Change-related risks include unintended consequences, negative business impact, compromise of information security and data loss or corruption.

Change-related risks also include failure to realize expected business value as a result of IT delivery delays, or implemented changes that don't produce the anticipated outcomes. A programme that focuses on the technical aspects of change at the expense of business risk considerations is not going to be successful; but it can be challenging for IT people to understand the relationship between the two.

Ironically, the goal is not zero risk but rather an acceptable balance of risk and desired business outcomes. Historically, change management sought to eliminate risk, often at the cost of delivery delays, but these days decisions to delay changes based on risk concerns must be carefully weighed by the business. In some cases, the business may choose to accept identified risks because of the significant window of opportunity afforded by timely delivery.

Example

A major corporation was pursuing an omnichannel distribution strategy, where in-store staff would pack and ship items ordered online via a sophisticated distributed inventory system and delivery logistics optimization.

Commitments were made to shareholders for omnichannel sales, but the timeframe for implementing the proposed system was tight, and the IT organization had to make use of existing warehouse applications to cobble together a shipping portal for store employees. The result of the rushed effort was an extremely 'unfriendly' user interface.

In traditional IT change management, the application would undoubtedly have been challenged by the change advisory board (CAB) and potentially sent back for usability issues to be fixed.

Nevertheless, the company released the application. Although it was difficult to use, shipments were made, error rates were tolerable and shareholder commitments met.

Was it the right decision? This organization thought so.

Management of risk is beyond the scope of this publication, but generally falls under the ISO 31000 standard. ISACA's Risk IT (included in the broader COBIT® framework) is an excellent reference for managing IT risk.

Information security management, which falls under the ISO 27001 standard and related frameworks (such as NIST cybersecurity framework), is also outside the scope of this publication.

No change management programme can guarantee zero risk; and this is not the goal of risk management. The appropriate management of risk strikes the optimal balance between risk elimination on the one hand and value realization on the other.

At a minimum, change management must ensure all changes have been adequately evaluated for risk. Keep in mind that many risks are not technical in nature, including:

- Delay in delivery of changes

- Failure to realize anticipated business value from changes

- Failure to meet windows of business opportunity which are dependent on changes.

Change management is a critical component of an organization's broader risk management programme, which must be able to demonstrate that changes are evaluated for risk, and that risks are documented, analysed and dealt with appropriately. Business management ultimately determines the organization's appetite for risk and acceptable strategies to manage it.

1.1.3 Minimize negative impact

Frequently referred to as 'avoiding unintended consequences', minimizing negative impact is perhaps the best-known objective of effective change management.

Unintended consequences are not limited to technical and infrastructure issues, incompatibility of components and the like. They might also include:

- Usability issues

- Degradation of performance of existing parts of the service

- Unexpected results in other areas of the service – for instance, the incorrect calculation of payments in an area of the application that 'wasn't part of the change'

- Other services on the infrastructure that no longer work correctly

- Other applications impacted by changes to underlying data and data structure

- New usage patterns that cause capacity issues

- Accidental compromise of information security

- Inability to use the changed system correctly because either support staff or users have received insufficient training

- Failure to update configuration information for this change that complicates the capability to make future changes or support this one

- Failure to notify appropriate stakeholders in a timely manner.

The IT infrastructure is a complex interrelated mixture of legacy applications and platforms, modern multitier application servers, both internally and externally hosted. For all the investments in fault tolerance and high availability, it's often the least obvious, simple things that cause the greatest disruption.

Even the smallest changes to a single component can have a surprising impact on seemingly unrelated systems and services.

Changes must be analysed to ensure there will be no unintended impact on the business. Configuration management is essential to understanding how components are interrelated and how changes to a component will impact on others.

Testing, especially system-wide regression testing and change modelling, can significantly reduce unintended consequences. Many organizations do only minimal change validation because of limited testing resources (and reliance on manual testing methods). Limited testing significantly reduces the ability of change management to effectively evaluate changes.

In order for change management to mature, you'll eventually need to invest in test automation. I'll talk further about testing and change validation later.

1.1.4 Business outcomes

'Outcome' is a word I've used a lot so far, and for good reason. Much of what has been done under the umbrella of traditional 'change management' has been process-focused (e.g. how IT can reduce the number of errors that get released into production). The outputs of the change process include the implementation of the required changes with some level of assurance that they won't have any negative impact.

Business outcomes are tangibly different in that they are closely related to customer satisfaction and the success of the business itself. Shifting the focus of change management

---

Business risk

A fast-paced start-up decided to upgrade its customer portal in the hope of increasing 'conversions', i.e. website visitors who complete a purchase during their visit.

A change was requested for a redesign of the web portal, based on industry best practices and customer feedback. IT and business collaboration was high as requirements were gathered and a design worked out.

There was much excitement as the release approached. The company had announced the new portal on social media, with great expectations.

The implementation was planned and executed well, and the IT staff celebrated their part in a successful change implementation.

When the conversion numbers were unchanged on the first day, the company assumed it was just visitors getting used to the new site. But after a month, the conversion rates were found to be lower than before the upgrade.

Was the change successful?

Well, that depends on whom you ask. The IT staff did an excellent job with the technical implementation. For its part, the business had done its homework and had designed an innovative solution.

So, whose fault is it that the outcomes were not achieved?

Remember, this is a start-up company, whose very survival depends on increasing sales.

The reality is that, in the end, both IT and the other parts of the business are in the same boat – which will either sink or float depending on the business outcomes they achieve.

from simply making sure new changes don't cause problems in the production environment to enabling desired business outcomes is critical to change management achieving its promised value.

To be successful, therefore, change management practitioners must first understand what outcomes the business is hoping to achieve. Changes must then be evaluated to determine whether they are likely to ensure that the desired outcomes are realized. Stakeholders must be engaged throughout the process, and results constantly evaluated against the required outcomes.

If your change programme has been primarily an internal IT effort focusing on the technological aspects of change, this is a significant shift that will require careful handling.

Implementing changes versus enabling outcomes

A critical issue for many organizations is to keep software up to date. Newer versions of office productivity suites, for example, provide additional features, eliminate bugs and maintain supportability.

Many organizations treat these updates as a change; after all, if the goal is just to get all the office PCs updated and doesn't consider the transition of users from the old version to the new, issues may well arise.

When the update is pushed out by a business division, and each installation is verified, there may be a tendency to believe that the change is complete and therefore a success.

However, after such an upgrade, if users are unable to open older file formats or have trouble accessing the application features they need to do their jobs, then the upgrade (change) has actually produced a negative impact on the business. Yes, the new suite is supported and has fewer bugs, but the outcome the business envisioned (stated or unstated) of users being able to continue to work with minimal disruption has not been realized.

Changes are necessary to support the business, both to keep the IT infrastructure stable and add/change/update business-required services and functionality.

Changes that will not be beneficial to the business should not be approved. This requires the IT organization to work very closely with the business. It also requires IT staff to use their knowledge of technology and infrastructure to assess how requested changes will help or hinder the business goals.

Change management roles in IT are quite challenging on account of this extra requirement, in addition to managing the technical risks and details, of being able to anticipate how changes will impact on users. This may require the development of training (for support staff, users and IT operations) and transition plans for both users and IT operations.

1.1.5 Governance and compliance

Increasingly, organizations face governance and compliance expectations around managing changes. These requirements are at the organizational level, and IT must ensure that requirements are consistently met. Change management serves as a critical control point where compliance objectives are enforced.

1.2 Change management versus organizational change management

The scope of this publication is IT change management – managing changes in the IT environment. Another term that causes some confusion is management of change, or organizational change management.

Organizational change management, which deals with managing people through transitions, is a broad topic that is beyond the scope of this publication but warrants a brief discussion because it is interrelated.

When the Greek philosopher Heraclitus declared that 'change is the only constant in life' he had no concept of the complexities of the modern world, but it turns out that his words were prophetic.

The focus of organizational change management is on the people side of change. Changes such as introducing a new product, or entering or divesting from a market have an impact on the people of the organization. Likewise, corporate reorganization or downsizing are significant changes. Even something as simple as introducing a new desktop office suite has a people side to it. All of these require both technical changes that must be managed and people-related issues that need a different kind of careful management.

As a practitioner, you cannot afford to ignore the people side of change management. This publication includes a fair amount of guidance for you as a practitioner to address this aspect while you work to adapt change management for your organization. I've found that people, and with them the organizational culture, comprise the largest challenge faced in building a successful change management capability.

The multiphased approach described in this publication is based on my experience of successfully introducing the concepts of change management while minimizing cultural resistance ('pushback').

So, while IT change management and organizational change management are two distinct topics, you'd be ill-advised to try to build change management without managing organizational change.

1.3 A brief history of change management

Gone (but not forgotten) are the days when changes were implemented as needed by technical experts who essentially owned the applications and infrastructure. Developers had complete control of the entire (mainframe) environment; changes went through a fairly rigid test process that accurately modelled the production environment.

These experts were revered with near deity status. Customers had no idea how any of it happened, but they knew it was complex, and were just happy when everything worked. With complete understanding and control of the entire environment, there were seldom any significant problems. When there were issues, the experts were available to fix whatever broke. Only limited communication was needed with others before making changes.

This created a wall of sorts between the all-knowing developers and support teams – such as service desk and operations staff. Support and operations generally had no advance warning that changes were planned or completed. Often the first clue for the support staff was provided by users calling the desk to find out what happened.

This attitude remains part of the IT culture; the only difference being that modern technical experts manage smaller pieces of technology. But as the IT infrastructure evolved from the mainframe and centralized data centres to a more distributed model, complexity increased significantly. With the increase in complexity, the likelihood of interactions between components adversely impacting on other parts of the infrastructure also increased, along with the corresponding diagnostic difficulties.

The communication and coordination issues in IT culture are difficult to overcome. One of the earliest forms of change management was simple and logical: have a meeting once a week with all the key players – development, operations, support, security – and talk about upcoming changes. The modern-day CAB evolved from these simple meetings.

Meanwhile, those working in the field of engineering, by virtue of its critical nature – designing and building things such as bridges and medical technology – have a clearer understanding of the need for precision in managing changes. It is undeniable; when lives depend on engineered systems, you cannot ethically just make changes and hope for the best: there is the need for more rigour and diligence to ensure changes that don't have unintended consequences.

Manufacturing engineering has rigid processes around the introduction of change orders for even the most minor changes. Great effort goes into keeping detailed records to track exactly when changes were made, and by whom. This enables quality engineering and field repair staff to know exactly how and when each product was manufactured and who to contact if warranted.

As IT change management evolved it adopted more from the world of engineering and the CAB became more structured and formal. Solid plans were required to help CAB members understand how the proposed change might impact on other areas. What parts are being changed, and in what ways? When and how will the changes be implemented?

It is important to understand something of the history of IT change management because, as the saying goes, 'Those who fail to learn from the mistakes of their predecessors are destined to repeat them.' (paraphrase, George Santayana)

1.4 Implementation challenges

Change management affects many people, both in IT and the business; there is no part of the organization that doesn't have a stake in it.

It is no surprise that this diverse group of stakeholders is the source of most of the challenges associated with building an effective change programme – they each have a unique interest in the process.

1.4.1 Prior efforts

If an organization didn't appropriately address the organizational change management issues associated with an earlier attempt to implement an IT change management capability, all future efforts are likely to suffer a negative impact. It doesn't really matter why the earlier effort was unsuccessful; the current effort will be impacted.

This could, in fact, be the single largest hurdle you'll face. Whether you're implementing the basic change programme I describe in Chapter 3, or maturing an existing one, the organization will remember and carry the baggage from previous efforts.

If you're new to the organization, the best course of action is to find out about any prior organizational efforts. Talk with staff about how they feel about IT change management; try to understand what worked and what didn't in any earlier endeavours. You need to know what people thought about the previous attempt, because, like it or not, your current effort will be viewed as another chapter in the previous effort.

Given that it's nearly impossible to distance your current effort from the organization's history, you'll want to look for ways to incorporate the learning from the earlier effort. If the previous effort was particularly painful, and people have strong feelings about it, consider a strategy to address the specific concerns raised. For example, if the previous effort had marathon CAB meetings, build the new change programme to keep CAB meetings short (I offer some strategies for optimization in Chapter 5). Make commitments to staff and follow through; give people every reason to believe that your current change programme will not have the same problems as the previous attempts.

1.4.2 Cultural concerns

Every organization has its own unique culture. Mature industries such as banking, gas and oil tend to have a slower pace and more rigid structures and policies (in some cases required by compliance oversight), whereas a Silicon Valley start-up will be much faster paced and more flexible. Government, military and healthcare are heavily regulated and policy-driven. Organizations that provide IT services to other companies will tend to have a more direct appreciation regarding why IT processes are important to their customers.

The list goes on, but you get the idea – each company culture is born out of the environment in which it operates, so it stands to reason that change management, or any new capability, must match the culture. This means you'll either have to work to change the culture (which is very difficult, time consuming and beyond the scope of this publication) or adapt what you do to fit into the existing one – which is greatly preferred, and much more likely to have positive results. Without a working knowledge of the culture, you can easily make assumptions and risk a negative reaction in unexpected ways.

A company that prides itself in being egalitarian and collaborative may view a simple concept such as process ownership as dictatorial and not a 'company value'. Likewise, trying to implement a best-practice process 'by the book' has its own challenges: highly innovative cultures may perceive 'best practices' as tantamount to mediocrity and a 'dumbing down' of their industry-leading ethos.

In practice, it's not nearly so black and white, and it's important to chart out a strategy that's clear and intentional about any cultural elements that need to change. Equally important is a plan to construct and communicate change management in ways that meet the key organization objectives, are culturally palatable and, most importantly, clearly articulate 'What's in it for me?' for each stakeholder impacted by the effort. This is the challenge for the ITSM practitioner.

1.5 Process versus outcomes

One of the reasons the approach in this publication is so successful is that it does not focus on the process. The thinking behind the process approach is sound – a consistent, efficient and effective process produces high-quality, reliable results. For many years, 'process' has been the antidote to inconsistent and/or poorly performing IT delivery.

Outputs versus outcomes

When I set out to purchase a new car, there's a series of things I need to do, such as establishing a budget, defining what I need, researching brands and models, test driving, selecting and purchasing a car, and driving it home. Though most of us are informal about it, the whole effort can be viewed as a process. The entire procedure involves inputs, activities and outputs, which sounds like the classic definition of a process.

When the process is complete, I have a new car. The output of the process is the acquisition of a new car; presumably one that meets the requirements (such as budget and features). But that's as far as the car buying process goes.

Because I bought the car, however, I'm now able to do some things that I couldn't do before – I can drive myself to work or take my family on holiday without using public transport. These are *outcomes* of the car-buying processes.

The outputs by themselves have very limited value. A car that sits in the garage because it doesn't have enough seats for my family can't help achieve the outcome of a summer beach holiday.

Processes produce outputs. Outcomes are how the outputs are used to provide value. To put this more succinctly in IT terms: a report is the generated output; how the information in the report is used to make decisions is the outcome.

There's nothing inherently wrong with process improvement. But let's be clear – if improving processes does not produce a commensurate increase in value realized by the business, not only does it lack positive value, but the perception of the effort will be negative. Period.

By contrast, the increase in value achieved through outcome-driven change management will be perceived as highly beneficial.

1.6 A clear business reason for IT change management

Any process improvement effort must begin with clear and communicated business outcomes. Let me be very blunt: implementing a change management process by itself is a weak rationale for investing in change management. While it may be exactly what's needed, a change management process is not in itself a valid business outcome.

My position on this isn't a simple matter of personal preference or opinion. While change management is one of the most frequently implemented service management processes, it is also the one with the strongest negative perception among IT staff.

The biggest challenges faced in building a successful change management capability are nearly all cultural – in other words, people-related: how people feel about change management, and whether they're willing to get on board.

This means it is critical to establish a clear business case for change management. It is important to address how staff will benefit from the new change capability. With human nature being what it is, people are reluctant to endure the transition until the benefits of the change itself become more appealing than the discomfort the change presents. If you lack a compelling case that staff can understand and buy into, you'll face everything from apathy to outright resistance.

Many organizations commission a change management programme on the heels of a high-impact change-related failure – often accompanied by significant negative financial or other business impact.

The good news is that this provides strong business case and senior IT management support for a change programme. They recognize a pain point and are convinced that improvements in change management are the answer. The case is clear: badly managed changes adversely impact on the business. The call often comes down from the chief information officer (CIO): 'I want change management, and I want it now.'

Significant improvement efforts that succeed typically have one feature in common: a critical and committed high level of senior management support.

The bad news is that the organization may not fully understand the magnitude of process and organizational changes required to adopt an effective IT change management capability. This is the classic case of mismatched expectations: leadership expecting one thing (immediate reduction in change-related issues), and change management implementers another (processes, metrics). It also suggests that one of the keys to success is both understanding and managing expectations – not just for management but for all stakeholders.

These reactionary programmes generally go astray in one of two ways: they try to implement either a comprehensive (and complex) change programme, often 'by the book', or an overly simplistic programme that doesn't produce the anticipated reduction in change-related failures.

In either case, senior management may lose interest long before the programme produces any tangible value for the business. In other words, without a strong and determined commitment to improving change management, the result is likely to be immediate fire-fighting. Patience has a tendency to run out, and with it go the resources and support needed to achieve meaningful results.

Every organization that is thinking about an IT change management programme (either a first-time implementation or maturing an existing one) must grapple with a few fundamental questions:

- Why do you want to invest in a change management programme?
- What results does the business expect?
- What level of capability maturity is required?
- What level of investment is necessary?

These questions warrant a fair amount of thought, as the answers will determine both your path and the difficulty in following it.

Some things to consider include:

- What problems do we experience?
- What do we need from change management that we currently don't get?
- Will improvement in change management be more valuable than other things we could be doing with same the resources?
- Does the business share the view that change management needs to improve? What problems do they see?
- Are the current problems exclusively change related or are there other contributing factors (e.g. IT strategy, design and architecture, development and testing)?
- Are there policy or regulatory requirements for change management that are not currently being met?
- Is the organization prepared to commit the required level of resources for the required length of time to ensure success?

● What is the downside of not investing in change management immediately?

● What is the organization's current state of change and/or transformation?

● Is now the right time to invest in change management?

Without a clear and compelling case to invest in change management, you're faced with a challenge. Any attempt to move forward with insufficient support for change management is to risk falling short of expectations or outright failure. I don't want to be dramatic about it, but there are lots of factors to consider and most of them have little to do with IT processes.

Let me also address a related issue. Where a change management improvement effort starts at the grassroots level in IT as a means of internal improvement, it may be seen by them as a self-evident need. It's true that incremental improvements can be made with limited involvement with the rest of the business, but, as I describe in Chapter 2, such an internally focused effort can unknowingly impede improvement in other areas.

Weak reasons to improve change management
There are numerous reasons for wanting to improve a change management capability. Some of those reasons are stronger than others.

Here are a few weak ones:

• To get people to follow the process

• Because the company I used to work for did it (or did it better)

• I just came back from training and I'm really excited about it

• To ensure all changes go through the CAB

• A best-practice framework says so.

1.7 Business value

The role of the board of directors, company president and the CEO is to ensure the resources of the organization are achieving maximum value for the stakeholders. At a very fundamental level, every asset of an organization exists for one purpose: to achieve the goals of the organization most effectively.

IT resources are no different. Companies invest in IT, staff and infrastructure to maximize the likelihood of most effectively satisfying the interests of their stakeholders. Organizational capabilities such as change management are assets as well. The investments made in staff, training, process improvement efforts and tools should result in building organizational capabilities – enhancing business performance and eliminating barriers to success.

Building an IT change management capability requires the investment of organizational time and resource. Organizational capabilities do not come free; there is always an opportunity cost – the organization could be doing something else with those resources.

Each organization must decide what level of investment in process maturity it believes will achieve optimal business results.

It is important because CIOs must justify why they are investing precious IT resources in building a change management capability rather than in other business projects and schemes that could be viewed as being of more direct value to the business.

If the investment in change management doesn't show tangible results (i.e. ones that can be quantified in business value terms) then it will be an uphill battle to persuade management to continue to make the investment.

It is therefore critically important that as you structure your change management plans, you show tangible results in the short term to maintain management support for continuing the effort. In *Leading Change* (1996), John P Kotter emphasizes the important of generating these 'quick wins' in order to both build and maintain support for the initiative. For the business, investing in change management is just like any other venture: it is evaluated by assessing the return on investment.

Businesses must optimize the whole, as a system, not just individual or specific parts. In the same way as marketing and manufacturing, IT is part of the business and should be funded at the level senior leadership believes will optimize the enterprise's overall results and deliver maximum value for the stakeholders.

This stands in contrast to IT organizations' self-balancing investments in internal efficiencies, business-focused projects and infrastructure readiness. The taking of resources away from business-focused projects must be kept to a minimum; IT capability investments must be weighed against all other organizational investments.

Keep in mind that the business is interested in improvements in change management (results) – not the implementation of a fantastically engineered best-practice change management process. Unfortunately, this is where too many well-intended change management efforts get shipwrecked. Given that the overarching goal of the change management effort is the value it's producing to the business, a basic process that demonstrates improved change management is better than an impressive process where there is no improvement in tangible results.

It is helpful at the outset to do a gap analysis of current change management versus desired end state. Identify the specific areas where the current change capability is working well and where improvements are required. With that in hand, determine the optimal target end goal and assess where the two differ. This will help establish specific areas where improvement is required as well as creating a clear end goal, which is critical in managing expectations.

Suffice it to say that investing in change management is much more than a simple operational decision. It is important to understand organizational challenges and constraints at the senior level. Appreciating the constraints will help you establish realistic and attainable maturity goals for your change effort as well as secure the required senior management support.

It is important to note too that 'business value' must be quantifiable and measurable and expressed in business terms. This is the guiding principle of all successful change management efforts.

1.8 Chapter 1: key concepts

The key concepts in Chapter 1 can be summarized as follows:

- Change management is a capability to:
  - Support timely and effective implementation of business-required changes
  - Manage risk appropriately
  - Minimize negative business impact
  - Ensure changes deliver desired business outcomes
  - Ensure governance and compliance expectations are met
- Focus on business outcomes, not process outputs
- Recognize that IT change management is not the same as organizational change management
- Start with a clear business reason for implementing (or maturing) change management
- Appreciate that the implementation of change management is greatly affected by the culture of the organization
- Remember that all IT-improvement investments, including change management, must be quantified in business terms to demonstrate business value.

2 A practical multiphased approach

It is not necessary to change; survival is not mandatory. W. Edwards Deming

2.1 The change management challenge

Aside from incident management, change management is arguably the most well-known and widely implemented IT service management (ITSM) process.

Unfortunately, many IT professionals have negative feelings about change management.

Why? I believe it comes from:

- Heavy reliance on centralized control
- Rigid focus on process (for the sake of process) compliance
- It being viewed as a quality inspection point
- There being little or no emphasis on the realization of business value.

There are two common approaches to implementing change management:

- Implement a very basic change programme that reviews all changes before release.
- Implement a fully mature change process as shown in best-practice frameworks.

By far the most common form of change management is the former – that of reviewing changes before they are approved for release into the production environment. This approach is primarily focused on the technical details of changes and serves as a gatekeeping function to protect the production environment. In this view, change management is all about analysing and avoiding any adverse impact caused by proposed changes.

This model inserts a change management checkpoint between a design/build and release into production (see Figure 2.1). When the work is completed and ready for release and deployment, change management is engaged as a simple last check to avoid any unintended technical impact.

Figure 2.1 Basic change management checkpoints

The good news is that this basic form of change management can and does do much to avoid some unintended consequences. However, it falls far short of the end-to-end control required to ensure the execution from approval to initial design through to final deployment produces the required outcomes.

By way of ironic contrast, W. Edwards Deming's revolutionary view, more than 40 years ago, was that quality should be engineered into the system/process rather than being inspected after the product has been built.

2.2 Starting small

Many organizations start with a very basic change management process that involves a programme of checking just before implementation. One of the main reasons for this is that it will be relatively easy to implement. Some of the other strengths and the weaknesses of starting this way are shown in Table 2.1.

Table 2.1 Starting small

Strengths	Weaknesses
Immediate results	Reactive only
Exposes changes to wider review	Limited ability to improve change during development
Minor change to workflow	Change not involved until after changes designed and built
Easy to implement	IT internal/operations focus

This idea is straightforward – get the right people together (i.e. the subject matter experts) and review all changes before proceeding into production, and by so doing, protect the production environment from unforeseen consequences. In this form, change management becomes a gateway through which all changes must pass. The goal is to reduce the likelihood of issues associated with the change under review.

Staff should understand that the point of the review is to avoid a repeat performance of a previous type of failure. The review also serves as a point of communication – where every team, including support teams, get the opportunity to examine and understand the change and ask questions.

Getting people together to review changes has value in that it should reduce the number of failed (or rolled back) changes by anticipating a number of unintended consequences. When the issues deal with minor details such as logistics – staffing, scheduling, business timing – the implementation plan can be adjusted to avoid them.

Unfortunately, without a broader picture of the purpose and value of change management, it can be viewed as an impediment to delivery that slows things down; at the same time it can demonstrate lack of trust in IT staff to do things right.

A last-minute, reactive-only change management can easily turn into a convenient defence if something happens to go wrong. 'Did it go through change management?' becomes the first question when a problem arises and the ability to answer, 'Yes it did,' becomes justification for poor planning and execution. This form of change management not only fails to ensure business outcomes, it is counterproductive and obstructionist. It also casts a negative image of change management in the minds of IT staff that is almost impossible to shake. Many organizations that have implemented this form change management continue to struggle with change-related issues.

It is at this point that organizations must make some tough decisions. It is important to recognize that adding more rigour to an unpopular change programme won't necessarily make it better. Very often the opposite occurs: the addition of restrictions and requirements typically produces staff resentment and resistance. Alternatively, some organizations decide to either abandon formal change management or allow the change advisory board (CAB) to become little more than a rubber stamp committee. Either approach has the same impact as adding more rigour.

This is one of the reasons organizations are desperately looking for more streamlined and efficient methods of managing change (such as DevOps and other Agile methods), which is unfortunate, because change management was never intended to be an obstruction.

The complete best-practice change process is designed to work throughout the lifecycle of a change (from the approval to design and build to the approval to release and deploy – and everything in between).

As you can see, although starting with a last-check change process makes a lot of sense and is easy to implement, it has a number of built-in constraints, its value is limited and it is very difficult to mature.

2.3 Quality inspection versus quality engineering

You may well be familiar with the work of W. Edwards Deming, the quality and systems architect whose landmark work with Japanese auto manufacturers in the 1950s and 60s formed the basis for modern quality engineering.

In *Out of the Crisis* (2000), Deming suggests a broader problem with the last-check approach:

> *Cease dependence on inspection to achieve quality. Eliminate the need for inspection on a mass basis by building quality into the product in the first place.*

Deming's point, in brief, is that finding defects earlier in the process is vastly preferable to discovery at the end. He suggested the best way to accomplish this was to engineer quality into the process.

The IT development process is surprisingly similar to traditional manufacturing. At each step of the process, work is performed, features are added and the solution is readied for eventual release to the customer. The aggregate time and resources invested increases at each successive step.

In manufacturing, if significant defects are discovered at the end, the product may be unsuitable for the customer. In the case of IT, if defects are discovered at the end of the development process, you're forced to:

- Accept it as is
- Attempt to minimize the impact and release it, or
- Send it back for rework.

None of these options are particularly good for the business; they increase costs, delay the realization of benefits or reduce anticipated features or capabilities (and the value they enable).

Reworking a change after it has been completed delays release and adds extral cost. If it can't be reworked, the change will have to be discarded.

Processes that rely on an inspection step at the end have limited value. In a practical sense, if significant issues are discovered during the inspection process, there are few options and all entail a less-than-optimal compromise to meet customer expectations – with the corresponding increased negative view of IT.

Issues that are identified by a last-check change process come at a cost – both in terms of delays to the business and the associated cost of remediation. A high-quality process where defects are found earlier results in less financial loss – even if the change is reworked or scrapped.

2.4 Starting big

On the other hand, trying to implement the fully mature process directly, as described in the best-practice frameworks – using a framework as a 'cookbook' or recipe source – is not only challenging for most organizations, it's also the wrong approach.

Many people attend training and certification courses in these frameworks, and the guidance given there is sound and encompasses best practices in use in organizations large and small all over the world. But if you're starting with little or no formal change management capability, it's a pretty big leap to go from an ad hoc process to one that is unified and mature. Table 2.2 shows some strengths and weaknesses of starting with a big change programme.

Table 2.2 Starting big

Strengths	Weaknesses
Comprehensive	Big leap for culture
Changes managed in the workstream	Requires significant resources to implement – and usually the application of organizational change management
Change involved at earliest stage; influences direction of change	Major change to workflow and organizational structure/culture
Business-outcome focused	Risk of losing management support

This approach initially makes a great deal of sense, and you can easily see how it addresses the very issues faced in your organization. But back in the real world, you'll face challenges that the books and training typically aren't intended to address. The course covers what to do, not how to do it at your organization.

The change management process described on these courses is complex and depends on other processes, such as release and deployment and configuration management. But many organizations start their ITSM programme with change management and without having these in place.

A large change management implementation takes resources away from traditional project work and usually won't demonstrate value quickly. This is where many well-intended change programmes earn a bad reputation.

Business and IT staff perceive change management as bureaucratic and slowing progress, creating resistance and ultimately undermining the success of the programme.

All this works against the success of a change programme. You risk losing support of the IT staff and the business.

2.5 Find the right balance

I've described two very common approaches to change management. In practice, these are just two ends of the spectrum. Every organization is unique, and every change management initiative is different.

Where these are helpful is in recognizing the trade-offs associated with either approach, and how that will work within your organization. Understanding both the culture of your organization and the desired end state for change management will help you structure your phased approach to best suit your situation.

For instance, if your organizational culture is rigid and structured by the nature of the business – healthcare, defence contracting and the like – it will tend to be more readily accepting of a more structured approach. However, if your organization is unstructured, heavily siloed or decentralized, taking smaller steps may be necessary.

In Chapters 3 and 4 I describe a balanced approach, starting with a basic phase followed by a second phase that adds to the change capability. I include a third phase in Chapter 5, designed to optimize a maturing programme.

In practice, it's the phased approach that's key to success – not rigidly following the phases as I describe here. Elements from later phases can be included in earlier phases, as needed, to address the particular challenges you face in your organization.

Much of the challenge in finding the right balance is adapting the best practices as needed to best fit your organization (see Chapter 6 for more on adopting and adapting best practices).

2.6 Cultural concerns

Every organization has a culture. Each is unique and influenced by the nature and challenges of the business, the attitudes of the staff and the behaviours commonly accepted.

As you begin to understand how the organization handles (or doesn't handle) changes, you'll learn a lot about how information flows through the organization. Be on the lookout for who is communicating what information with whom. This will give you tremendous insight into how various types of change are dealt with.

Recognize that as you introduce change management, you may challenge some very basic cultural values, one of which may be that IT staff were empowered to manage their own changes in the past – self-contained within IT.

Formal change control can feel like the loss of autonomy. Staff may feel they are no longer trusted to do what they've been doing for years. This may be especially true if your change management programme is viewed as a purely reactive response to some high-visibility incidents or failures. This makes it important to engage with all those affected to ensure they understand the intent and the justification – hopefully in the context of their own self-interest.

Organizations just getting started with formal change management will undoubtedly face resistance, because there will be significant changes for nearly all IT staff, and for many in other parts of the business as well. How it is handled will have a significant impact on the programme's likelihood of success.

In their seminal work, *ABC of ICT: An Introduction* (Wilkinson and Schilt, 2008), Paul Wilkinson and Jan Schilt put it this way:

The next conference you attend, look at the program and the titles of the sessions. 95% will focus on some framework or method or approach or specific process, very few will focus on addressing attitude, behavior and culture, telling you how to embed the solution in the organization, what sort of resistance you will encounter and how to overcome this. This leaves many organizations, new to adopting frameworks, with the naïve belief that they can simply be 'implemented'.

The ABCs they describe are attitudes, behaviours, and culture. Collectively, these represent the 'people side' of process adoption. By comparison, the processes themselves are actually straightforward.

One of the unfortunate downsides of traditional best-practice training is that the well-designed processes seem ready for immediate implementation. Many well-intended practitioners come back from training ready to implement a process, with no concern for cultural implications at all, which is unfortunate, because processes require people to be successful. Any effort to change or implement new processes invariably puts you face to face with organizational change.

Attitudes surrounding change management can touch on deeply held beliefs and cultural values such as:

- We hire top talent who know what they're doing.

- We expect our staff to do the right thing.

- We empower staff to make necessary changes.

- We trust each other to work independently, to do what's best for the customer.

You would be wise to take the time to understand how the new change management programme will fit within the organizational culture. Be very intentional about managing the cultural change that goes along with the new process.

Cultural objections commonly raised while implementing change management include:

- You say we're doing everything wrong.

- Why do other technical areas have to review/approve my changes?

- What we've been doing is working fine; why change?

- It won't work here.

- Best practices can only make you average.

Humans are complex creatures whose behaviours are driven by some deeply rooted needs and motives. Something as simple as implementing a basic change process can trigger some surprising Maslovian-like responses that can be very difficult to overcome.

When 'process-only' efforts struggle or fail, as they do, all too frequently, the first response is often to blame the framework. I've heard many IT professionals declare that 'we tried such-and-such a framework, and it doesn't work', when, in fact, the framework itself is fine; the problem is the unwise emphasis on the process to the exclusion of adaptation to the culture.

Faced with a struggling process-focused programme, the organizations sometimes resort to forced adoption of heavy-handed process compliance. All this does is create even more ill will, resentment and resistance. In other words, the net impact becomes worse than the original failed attempt. Sadly, companies that have gone down this path will forever have a bad taste of 'change management', making it all that much harder to ever build a successful change management capability.

The approach adopted for this publication places a premium on achieving a value-producing change management capability, and includes addressing the organizational change management required for success.

2.7 Keep it simple

Simplicity is one of the guiding principles of successful change management, especially in the early steps. Include only those elements that are absolutely necessary, and whose value can be immediately understood.

This is where the multiphased approach comes in. The approach seeks to maximize the immediate value of a very basic change programme, while setting the stage for successfully maturing the process as specific milestones are achieved.

2.8 A word about tools

There are several commercial ITSM tools that include support for change management. The selection of a tool is beyond the scope of this publication for a specific reason – it is impossible to choose or configure the right tool unless you understand change management and the unique needs and challenges of your organization.

I don't know who first said it, but the phrase, 'a fool with a tool is still a fool' pretty much summarizes my view on the subject.

I'm not suggesting that people or organizations with ITSM tools are in any way foolish. On the contrary – it is very difficult to have a mature change management capability without good tools (and there are a lot of good ones available), but change management, like other capability development efforts, depends on a balanced combination of people, process, partners and products (tools). As I've said, change management is not primarily a process effort – it involves the other elements in the right balance, as shown in Figure 2.2.

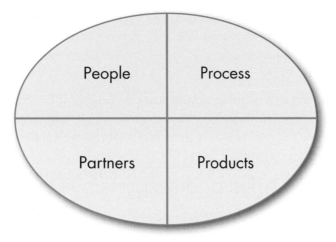

Figure 2.2 Elements of effective change management

I've seen far too many change management efforts that are little more than total reliance on a software tool without ever addressing the more significant challenges of adapting the principles of change management to meet the needs of the organization.

It's very tempting for an organization that's just getting started to look for tools to help implement a standard change management process. This thinking typically results in an attempt to use best-practice guidance as a template to be religiously followed without the needed starting point: people. It places tool and process before people, and the end result is failure. Rather than accept responsibility for the faulty approach, the blame is assigned to the framework: 'We did exactly what it said, and it doesn't work.'

The other typical thought pattern suggests that if the tool embodies the best practices, then implementing the tool 'out of the box' provides a standard process. Sadly, the idea of a standard or 'one-size-fits-all' implementation of change management doesn't exist, regardless of whether it is sourced from official publications or by installation of a tool.

There is no shortcut for the adaptation work you must do to make the process fit into your organization, and no justification for building your programme around a tool.

If your company already has an ITSM tool in place, I strongly encourage keeping your process development work separate from any tools implementation. Before you invest effort into the ITSM tool for change management, work through the people and process aspects. Use only basic tools, such as a simple spreadsheet. This will make it easier to keep your efforts focused on achieving the results the business requires instead of getting distracted with the specifics of how the tool should be configured. This approach will also keep you from including limitations of the tool in your process design. Remember, the goal is a change management process that delivers the expected outcomes and business value, not conformance to a tool.

You will learn a lot about how change management needs to work in your organization, especially in the first phase. You want the flexibility to be able to rapidly adapt your approach, so avoiding the unnecessary complexity of even a comprehensive tool is important for success.

Once you have a solid understanding of how your change programme needs to work, selecting and/or configuring a tool becomes a relatively simple task to meet the needs of your organization.

2.9 Clarity in roles and responsibilities

The introduction of formal change management will invariably necessitate establishing new roles as well as changing some existing ones. Having developed a solid understanding of how changes are currently managed, you should have a fairly clear picture of existing roles relating to managing changes.

Where roles need to be adjusted, there's an elevated need for clear communication to both the people affected and those who understand their previous role. This is where the RACI (responsible, accountable, consulted, informed) chart can be helpful.

How to make a RACI chart

A RACI chart is a table of rows for each task/assignment, and a series of columns, one for each stakeholder. At the intersections of the tasks and stakeholder, assign one of the roles listed in Table 2.3.

The only hard and fast rule for role assignment is that each task can only have one accountable role. This shouldn't be ignored or taken lightly. Discussion about who's ultimately accountable can reveal differences of opinion (potential sources of confusion) and begin the process of creating clarity.

If you struggle to agree on just one accountable person, you may have your tasks at too high a level. Break tasks down further until a logical accountable person for each becomes clear.

Next, identify who's responsible for accomplishing the task or assignment.

It is common to have more than one responsible role. But if you find yourself assigning lots of 'responsible' people, your tasks may be too low level.

'Consulted' can be a tricky role. Consulted means having critical knowledge, experience, or information necessary to successfully carry out the task, but does not actually do the work of the task. Be sure to keep a clear distinction between responsible and consulted.

Roles are not cast in stone. With the agreement of the stakeholders, they can be adjusted as needed. RACI charts are a perfect reference point when disagreements arise and they help facilitate healthy dialogue. In this first stage the RACI chart can be very useful in identifying disparate understandings of roles and responsibilities. The chart itself is less important than the clarity it helps bring.

Table 2.3 RACI roles

Responsible	The person who actually carries out the process or task assignment
	Responsible for getting the job done
Accountable	The person who is ultimately accountable for the process or task being completed appropriately
	Responsible people are accountable to this person
Consulted	People who are not directly involved with carrying out the task, but who are consulted
	May be a stakeholder or subject matter expert
Informed	Those who receive output from the process or task, or who have a need to stay informed

It is not uncommon for a basic RACI chart to be included in the change policy, but generally only at a very high level. For instance, it might say that the service desk manager is responsible for ensuring all approved changes are communicated with stakeholders. It is best to not use proper names (just roles) in a policy document.

2.10 The multiphased approach

With the perils of an overly simplistic approach (see section 2.2) on the one hand, and the challenges of the overly complex (see section 2.4) on the other, I advocate what I call a practical 'multiphased' approach in the following chapters.

Like many things in life, success in the earlier steps generates support for successive ones. Trying to adopt a fully mature change programme is somewhat like trying to run a marathon without having put in the time to build the physical conditioning and endurance. While it may be possible for some to just go out one day and run 26.2 miles, no one would recommend it as a good approach, and certainly not as 'best practice'.

The successful change management programme must demonstrate measurable results at each step along the way. With this approach, senior management, business leaders and stakeholders will see immediate improvement in change-related issues, downtime and delays.

Like the would-be marathon runner, we must take one step at a time. Once we're succeeding in the first, we take a second, and so on towards a change management capability that meets the current and future needs of the organization.

In the next chapter, we'll take a look at that first step (phase 1).

2.11 Chapter 2: key concepts

The key concepts in Chapter 2 can be summarized as follows:

- Starting change management either too big or too small has challenges.

- The change advisory board as a quality control inspection has limited value.

- Quality must be engineered into the development process (not inspected in after development).

- Success depends on mindful attention being paid to the attitudes, behaviours and culture of the people within the organization.

- Start small, and keep it simple.

- Don't be a 'fool with a tool' – understand your organization and the underlying change management needs first.

- Create clarity in roles and responsibilities (use a RACI chart).

- Use multiple phases for a successful change programme.

3 Basic change management (phase 1)

> *"The journey of a thousand miles begins with one step [in the right direction]* Lao Tzu *"*

In this first phase, we'll be taking our first step. In order to ensure our first step is, as Tzu implies, 'in the right direction', we must first take an inventory of how change management is currently being handled.

Formally done, this is what's commonly called a 'gap analysis' – an inventory of current state of change management along with a reasoned declaration of desired end state. Whether undertaken formally, or informally, it's important to have a good understanding of the current and desired future state of change management, as I'll explore in this chapter.

In the last chapter I talked about finding the right balance between attempting a too-large change management implementation on the one hand and a too-small one on the other. In this chapter, we'll examine the first of two phases scoped to ensure success.

If your organization is like many others, a basic change programme to review changes before release into production may already be in place. In that case, you should treat this chapter as a review to ensure you've got the basics covered before you move on to the next phase. Pay special attention to the success criteria to assess when you're ready to proceed further. It may be that the current process is lacking some of the essential elements I describe here. As I believe they are critical for successful maturity, I would recommend an initial phase that ensures all the elements described in this chapter are in place before moving on to Chapter 4.

Far too many change management programmes begin with a primary focus on process, instead of on people and related issues. Unfortunately, experience shows time and again that the process, by comparison, is relatively easy. As mentioned in the previous chapters, the real challenges are the cultural issues you'll face and being able to produce tangible results in a timely fashion.

It's worth noting that the concept of continual improvement is implicit in this publication. Any capability that doesn't include continual improvement as a core element is one that is destined to fail to meet future needs. Change management must be designed to be adaptive.

The focus for this first phase is to establish a successful basic change management programme, one that lays a solid foundation that can easily be matured to fit the changing needs of the organization (more on this aspect in the next chapter). This approach avoids unnecessary complexity, minimizes cultural resistance, and focuses on the business outcomes that change management can deliver.

3.1 Phase 1 goals

Because the concepts and ideas embodied by 'change management' mean different things to different people, it is vital to be very clear about what we are trying to accomplish in this first stage. If you don't carefully establish and manage expectations, whatever you do will be viewed by some people as a failed attempt, which places your new change programme at unnecessary risk, right from the beginning.

The goals for the first phase are selected to maximize the chances of success and minimize the difficulties in adoption.

The first phase goals:

- Build a culturally relevant case for change management

    - The importance of change management

    - What's in it for the stakeholders

- Introduce the concept of change control

- Show early results, demonstrating value to the business

    - Reduction in change-related incidents

    - Increase in successful change implementations

- Ensure cultural adoption

- Manage the IT environment while minimizing:

    - Bureaucracy

    - Unnecessary delays

    - Organizational resistance

- Set the stage for future maturity.

3.1.1 A case for change management

I've said a lot about people and culture being the biggest challenge you'll face when starting a change management programme. Not surprisingly, then, we start with getting people on board.

We start this process by creating and communicating a clear case for change management, by explaining why, exactly, the organization needs to invest in change management. The case should address key concerns such as:

- What's wrong with what we have been doing?

- Why do we need to do this now?

- How will the change programme make things easier or better for me and my team?

- What will the organization gain from formalizing change management?

The case must be consistent with management's expectations for the change management effort. Establishing clear senior management support for (and expectation of) change management will go a long way towards setting the proper tone for the upcoming programme. Anything less, as I've said, is to risk failure.

It is important to take the time to personally deliver the change management message. One on one with key stakeholders is ideal, In addition, it is critical that your message for change management is clear and consistent, and it must also be genuine. If IT staff perceive that you're just doing someone's bidding, or don't have a real passion for it yourself, the message will be clear – the effort really isn't that important, and their support is optional.

These conversations should be two way. Take the time to communicate the change management message, but also take the time to listen to other people's thoughts and opinions on the matter. Include their input in the planning process as much as practical.

The importance of getting the support of IT staff cannot be overestimated, especially in the first phase.

3.1.2 Introducing change control

The idea that change should be controlled is not new. Changes are a major source of incidents and (negative) business impact that is almost entirely under the control of the people charged with design through to deployment. The ability to anticipate and avoid negative business impact has always been part of good operational practice.

For many years, internal and external IT audits have been concerned with ensuring that changes are properly requested, documented, approved and implemented to ensure no possibility of unauthorized changes. In addition, and irrespective of the type of organization (e.g. banking, government, retail, aeronautics), there will almost certainly be regulatory requirements that are already being taken into consideration.

One of the concepts of change control is the idea that all changes must flow logically from business requirements in support of business objectives. This means that change management is responsible for ensuring the alignment of changes with those objectives. This issue falls under the umbrella of IT governance, which is outside the scope of this publication, but I include key elements as they relate to change management.

If there is only a technical view of change management, the governance aspects of change management remain unaddressed. So, while changes may be effectively implemented with no apparent negative impact, the organization has no way of demonstrating that each change is appropriately authorized and supports defined (and documented) business objectives.

I'll cover change policy later, but it's important for management to clearly communicate their expectations for change management – especially the governance aspects that may not be self-evident to technical staff. With a clear change policy, then, IT management can effectively execute change at the operational level in support of policy expectations.

Every IT organization has some form of change management. There's no such thing as an IT team that doesn't implement changes that are managed in some way – ad hoc or formal – with varying degrees of success.

One of the key objectives in the first phase is to successfully introduce the concept of change control.

Change control includes these key aspects:

- Operational effectiveness
  - Risk management
    - Negative impact reduction
    - Business risk
  - Cost optimization
    - Rework reduction
    - Business impact reduction
- Communication and coordination
- Governance
- Achievement of business outcomes
- Time to value
- Compliance (regulatory and otherwise).

If your organization has little or no formal change management in place, some of these aspects are probably not addressed effectively. For instance, technical staff may be used to managing their own changes, and while they may recognize the need to be effective in managing the operational aspects of change (technical details, prevention of interruption to services etc.), they may not perceive the relevance of the change governance and regulatory compliance that comes with formal change management. This is one of the major challenges you'll face.

From a technical viewpoint of change management (almost as an extension of software testing and quality assurance), it's easy to see how the shift to a more formal approach and introduction of additional oversight could be perceived as criticism of staff – that they are doing a bad job and so now changes need to be reviewed by others from outside IT. This unintended insult is usually the initial source of cultural resistance. The solution is to engage stakeholders in order to encourage them to suggest a wider range of reviewers; engagement is an important aspect of establishing and managing expectations. You need these very staff to be on board with your change management efforts; alienating them at the very beginning is counterproductive.

It is important to develop a clear understanding of the organization's current view of change control. This allows you to establish a plan based on the gap analysis between the current and desired states. This approach supports the creation of an implementation strategy that leverages the current capabilities and success of the teams while carefully introducing new concepts.

Take the time to ask, and understand the answers to, the following questions:

- Who authorizes changes to IT services?
- Are changes reviewed, and if so, by whom?
  - Peers
  - Technical leads
  - Unit managers
- Who is accountable for the success of individual changes?
- How are changes coordinated (or not) among the various technology teams (infrastructure, application development etc.)?
- How are changes communicated and coordinated with customers?
- Is there a project management capability? To what degree does it manage the project-related aspects of change management?
  - Portfolio management and optimization
  - Business prioritization
  - Clearly understanding desired business outcomes
- Does the organization's incident management process track change-related incidents?

In addition, it is important to understand some of the cultural elements that interplay with the introduction of change control:

- Teamwork and communication culture
  - Siloed versus collaborative
  - Us versus them
  - Open communication across organizational boundaries
- Risk culture
  - Organizational appetite for risk

- Calculated risks encouraged to achieve organizational goals

- Balance versus elimination

- Fixing blame versus fixing issues

- Empowerment versus avoidance of blame

- Leadership culture

  - Role and competency of leadership and management

  - Command and control versus empowerment

  - Policy compliance versus strategy and goals.

All of these factors come into play as you implement change management. It's impossible for me to give specific advice regarding how to address all the cultural issues you'll encounter as there isn't a single approach that will work in every situation. However, one thing I can assure you: you will face these and other issues. It would be unwise to proceed without taking stock of this fact.

The opposite of controlling changes is letting changes control us (or more correctly, controlling the production environment). This puts not just IT but the entire organization at great risk.

3.1.3 Showing early results (business value)

Many change management efforts begin with strong organizational support, starting with the senior management who are convinced that more effective change management will increase customer satisfaction with IT services. This is a great place to start, but unfortunately this initial support will fade if the programme doesn't show improved results, quickly.

The typical process-focused change implementation frequently includes a lengthy analysis and requirements gathering phase that often exceeds the management's timeframe for demonstrating concrete results.

Senior IT management face tremendous pressure to deliver business results. If your new change management effort doesn't show tangible improvement in change-related incidents in the short term, you risk the loss of support; management is likely to be forced to move resources to other more pressing business-value-generating activities.

The challenge is to produce tangible results quickly, at the same time laying a solid foundation to use as a basis for further improvement. Use the concepts of continual service improvement and 'Agile' development methodologies as part of your critical strategy for sustained change management support and success.

In sum: start small, show results. This builds support and momentum that is critical to maintain both IT staff and management's support.

3.1.4 Cultural adoption

As I mentioned previously, crafting a process to manage IT changes is relatively easy. There are many sources for 'standard' change management processes – including the various best-practice volumes; I strongly endorse these as excellent sources of process information.

But these frameworks don't address the people side nor the organizational issues of adapting the material to fit the organization. Adopting the processes as recorded in the framework ignores the recommendation to adapt (also included in the framework) in order to fit the business needs. Adopting is easy; adapting is the hard part – and is the source of most failures to implement a change management process that works and can be matured to fit the changing needs of the enterprise. Cultural issues are nearly always responsible for disappointing results.

Ignoring the organizational culture, and how the new change management programme will fit into it, puts your efforts of great risk.

3.1.5 Manage the IT environment

Given the complexity of the modern IT environment, if there is no effective change management capability, business outcomes and value are at risk. Minor changes in one seemingly isolated system can have far-reaching negative impacts in other areas and in ways that are hard to predict.

One of the most common reasons to add or improve a change management programme is to get a better handle on what's happening in the production environment, with the specific goal of reducing both the frequency and impact of change-related incidents.

This first phase provides guidance to allow you to identify and put the right controls in place to proactively manage stability and uptime and minimize adverse impacts on the business.

3.1.6 Future success

Most organizations will eventually need a mature change management capability. If the early efforts are rocky or disruptive, it will make it difficult (if not impossible) to get the required support for future improvements. The same is true if the organization views change management as only a technical last check to avoid failures.

The method in this publication puts a premium on not just the establishment of a basic change programme but the successful laying down of a solid foundation to support additional maturity based on experience and changing business needs.

If you start small, show early results and address cultural issues, you are more likely to maximize both the immediate and long-term chances for change management success.

3.2 Basic change management process

One thing that is common among all IT organizations, from the one-person shop to the major multinational corporation: they have some form of process for application/service development and another for release and deployment into the production IT environments, where they're used by people to do the work of the company. How these changes are implemented is at the heart of basic IT change management.

Without formalized change management, it is not uncommon for different development teams to have different practices around how changes are moved into production. Some organizations have a well-designed development and test environment to promote changes to production while others may develop changes in production itself. Some teams may have highly structured processes for handling production changes while other teams are completely ad hoc. Infrastructure teams – especially those responsible for data centres and other highly critical areas – often have sophisticated controls in place to minimize business impact.

Your organization may have haphazard and highly inconsistent change processes, and that's the point – you must first understand how changes are currently being implemented in your environment before introducing something new. It isn't safe to assume that all changes are implemented in the same way, with the same degree of diligence and rigour (or the lack thereof.) Take the time to understand how your organization handles changes.

Regardless of how new or changed services are developed and released, the most basic change management step is to insert a review after the development work is complete and before the change is released into the production environment.

This review step is the beginning of formal change control and starts with the formation of a change advisory board (CAB) invested with the review authority. It is, in essence, the addition of a quality inspection point.

This is a critical first step in introducing formal change management. All changes should be properly planned, built and tested before being reviewed by the CAB.

For many organizations, the CAB is the first time a cross-functional group will come together to review changes.

Figure 3.1 shows more details of the basic process flow, and how the various roles work together.

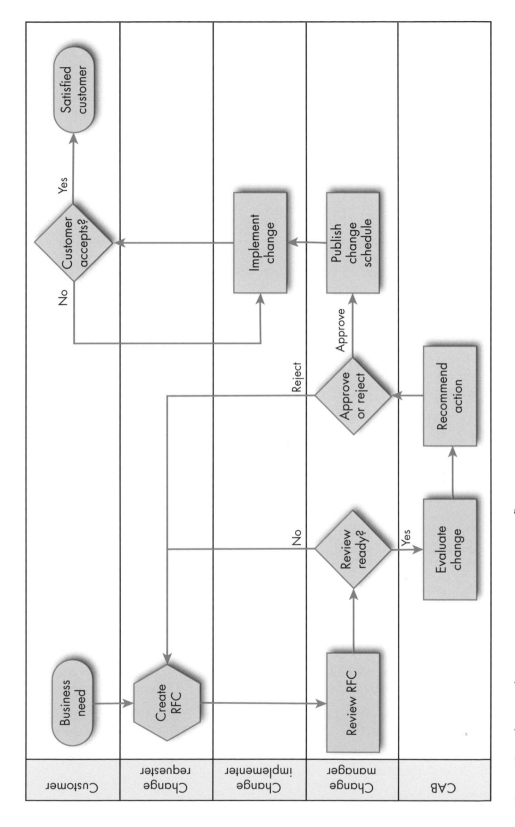

Figure 3.1 Phase 1 change management process flow

Changes start with a business need

A change always starts with a business need. This is fundamental for a change programme that is focused on business value. In practice, of course, many changes arise from within the IT organization. This is normal and expected, but the point here is that every change must be viewed from the standpoint of the customer. This is not a minor semantic – all changes are done for reasons that support business outcomes. To effectively manage changes, we must understand what the intended outcomes are, keeping in mind that valid outcomes can include 'continued stability', 'upgrade to maintain supportability', 'addressing capacity issues' and the like.

The view of change management as primarily an internal IT activity that only deals with the technical details of changes has sabotaged many well-intended change efforts. A change programme with this viewpoint struggles to mature beyond a very basic level.

To get started on the right path, change the approach to view changes from the standpoint of their value to the business as early as possible in your change programme.

The following is a description of how the process proceeds.

Once a need for change has been determined, the change process should be initiated by submission of a completed request for change (RFC) to the change manager, who should review it to determine whether it's ready for a CAB review.

Once the RFC is judged to be ready, the CAB evaluates the request and recommends action to the change manager. The change manager typically follows the CAB's recommendation, though there may be cases where changes require approval by a higher authority for that type of change or level of risk.

Approved changes are put on the change schedule as approved and the requesters and implementers are notified of the authorization to proceed.

After implementation begins and before the work on the change is considered complete, every change must be fully tested, and accepted by the customer.

Post-implementation testing is designed to ensure that the change achieves the desired results and without unintended consequences. Make the definition of test plans part of the change request. Any parts of the system that might be impacted should be tested as well. Full system regression testing is ideal, though not always practical or possible. If your organization lacks a formal test capability, some form of elementary testing must be performed to ensure change success.

Note that not all changes require formal customer approval – but at a minimum, all changes must be evaluated to ensure customers are not adversely impacted. For instance, if the service performs poorly after the change, the change should not be considered successful.

The concept of formal customer sign-off will be new to many organizations. For larger changes, project management handles customer acceptance. At this stage, it is enough to recognize that a change is deemed to be successful when the stated business outcomes are achieved – not when the technical implementation is complete.

The key steps are explored in more detail below.

3.2.1 Change policy

Because of the varied understanding of the practices that are part of change management, it is important to develop a clear and consistent picture of company expectations before making changes to existing practices. This is done with the creation of a change policy (a clear and concise statement of expectations, approved by management) that underpins the governance of the new change management approach.

Developing a change policy is important to create clarity of purpose and set expectations. The policy should clearly state management intentions around how change management will work. The policy documents what management expects to achieve from the change management effort.

The tone and formality of the policy should be consistent with the company practices. It needn't be lengthy or complex; in fact, every effort should be made to keep it from being heavy-handed and consequences-oriented, especially in the first phase.

It should address these key elements:

- Objectives of IT change management
- Scope of changes to be managed (what changes are subject to change management)
- High-level process flow
- Agreed definitions of the types of change:
  - Normal
  - Standard
  - Emergency
- Establishment of roles and responsibilities, and drawing up of a RACI chart (change manager, CAB etc.)
- Expectations for compliance (what happens if change management is bypassed?)
- Establishment of change windows and how exceptions are handled
- Critical success factors (CSFs; two or three at most) and key performance indicators (KPIs) to support each CSF
- Determination of key metrics to measure and report on the performance of the change management process.

Appendix A shows a sample change policy. It should not be used as is, because each organization is unique, and the policy must reflect the needs and expectations of your organization.

Communication is key to a successful change programme, and a clear policy statement is key to communicating management expectations.

3.2.2 Requests for change

The change management process begins with the creation and submission of an RFC. This is how change requesters formally notify change management of the need for a change. It's called a request for change because it requires review before being approved (or rejected) by the change manager. This may seem like a trivial detail but it can be the source of resistance if the staff were previously empowered to make changes without any oversight and now must make a formal request that could be denied.

Appendix B gives an example of typical fields required for an RFC. The fields vary from organization to organization depending on local needs. To avoid the CAB becoming (or remaining) internally focused on technical details, it is important that every RFC explicitly contains a section (or statement) that addresses the reason for the change which should describe the expected business outcomes.

Only include information in the RFC that is necessary for the decision-making process (including the risks if the change is approved and if it isn't). Don't include information that isn't relevant to the making of the decision. Additional fields should be added only as specific needs are identified and their use is deemed helpful in the process of reviewing changes more smoothly.

The RFC is submitted to the change manager. I prefer to use simple methods – a spreadsheet for example – because they can be readily adapted as change management matures. This helps keep the focus off the tool and on the business of effectively managing changes. In the case of the spreadsheet, the requester makes an entry on the spreadsheet and notifies the change manager of the submission. When the change manager believes that the RFC is ready for review, they will notify the requester and put it on the agenda for the next CAB meeting.

RFCs can be raised for a variety of reasons, including:

- Projects for new or changed services

- Issues arising from daily operations (incidents)

- Normal changes – maintenance and upgrades

- Service improvement plans (often associated with continual service improvements efforts).

As the RFC progresses, the change manager is responsible for ensuring the status is updated and communicated to the appropriate stakeholders. Examples of possible status updates to be included in the workflow document are: submitted, pending, under review, approved, and implemented. The pending RFC list, filtered by the review date, becomes the agenda for that CAB date.

3.2.3 Change types

Changes should be categorized by type, which helps to ensure proper handling.

There are three main types of change request, as shown in Table 3.1. As mentioned above, the change policy should include these definitions along with high-level process expectations.

Table 3.1 Types of change request

Change type	Description
Normal	Normal changes require a normal CAB review and are approved by the change manager upon advice from the CAB. Most changes that come to the CAB are of this type.
Standard	Standard changes are changes that are low risk, are performed frequently in daily operations, and follow a documented process that's been previously reviewed by the CAB and pre-approved by the change manager (see Chapter 5 for more details).
Emergency	The emergency change type is reserved for those situations where there is a significant risk that must be immediately addressed to avoid dire consequences (see section 3.2.6.6).

3.2.4 Change priority

RFCs should be prioritized based on business needs, not technical capability. The purpose is to ensure the changes with the highest priority to the organization are considered before changes with a lower priority. Priority should be based on both business urgency and business impact and aligned with the organizational governance.

Table 3.2 gives an example prioritization schema. There is no single correct schema for prioritization; however, part of the change evaluation process should validate and establish priorities for each change. Regardless of the particular schema chosen, the objective is to align change efforts with the priorities of the organization.

Table 3.2 Change priorities

Change priority	Description
Urgent	Reserved for changes of the most critical nature to the company
High	Changes that are causing severe impact to a significant percentage of users, or having an impact on critical business functionality
Medium	Impact is not critical, but the change cannot wait until the next scheduled release
Low	Change is required, but is low enough that it can be included in the next scheduled release

RFC submitters should assign an initial priority based on their analysis of the business urgency and the business impact of the change. The CAB must validate the initial priority and weigh it both against the other RFCs in the works and against the organization's priorities and governance processes.

3.2.5 Roles

Formal change management requires clearly understood roles and responsibilities. This is usually done in the form of a RACI chart.

Table 3.3 describes the key roles and a summary description of the associated responsibilities.

Table 3.3 Change management roles

Role	Description
Change management process owner	Ultimately accountable for the change management process itself (i.e. the design, implementation, governance and ongoing improvement). The role includes the development and maintenance of process documentation and ensuring the documented process is followed in practice.
	In IT service management (ITSM), all processes have a process owner.
Change manager	Responsible for carrying out the day-to-day tasks required to administer the process according to the documented policy. In smaller organizations, this role is often combined with that of the process owner and carried out by a single person. In larger organizations, there may be several change managers, each with responsibility for a geographic region, technical domain or other logical division of responsibility that makes organizational sense.
	Regardless of the structure, it should be clear that a change manager has both ownership and accountability for the changes coming through the change process.
CAB	Works as a high-level group to review proposed changes, including analysing risks and potential negative impacts, in order to provide the change manager(s) with expert advice on whether particular RFCs should be approved or rejected.
CAB member (see section 3.2.6.2)	Uses their knowledge and experience to evaluate RFCs. Consults with their team to ensure they are fully briefed on the implications of a change before CAB meetings.
Change requester	Documents the reason for and expected outcomes of a proposed change. Completes and submits the RFC form. Answers questions from the change manager and CAB – and may attend the CAB meeting to represent the change.
Change implementer	Responsible for the actual implementation of the proposed change. If the change is large or complex, the change implementer is often a senior technical staff member, operations manager or project resource.

3.2.6 Change advisory board

The CAB is the heart of basic change management and its creation is the beginning of formally managing changes.

This group is frequently misunderstood to be the change approval or authorization board. It's a distinction that's important to get right from the beginning: the CAB is *advisory*. This is because its job is to review proposed changes and advise the change manager of the results of its findings. It's the change manager who ultimately approves (or rejects) the changes.

Is the CAB a voting body?

Some organizations strongly embrace the concept of committees, which often includes having formal chairmanship and voting members. For such organizations, it's natural for the CAB to employ this kind of structure.

What results is a change process that is highly focused on the CAB and its regular (typically weekly) meeting that is chaired by the change manager.

I strongly advise against implementing the CAB in this fashion. This type of committee approach tends to establish rigid and inflexible authority centred on the CAB itself.

As your change management programme matures, the CAB should be increasingly concerned with the end-to-end quality of the change capability and less about individual changes. The role of the CAB is to ensure that the quality of all changes meets the needs of the organization. This places the focus on the outcomes of change management, not the process, and most definitely not on the CAB itself.

Because it is so key to a successful early adoption, let's take a close look at the CAB.

3.2.6.1 CAB role

'CAB' and 'change management' are not the same. Change management is the broader capability that manages the entire process of raising, reviewing, evaluating, approving, tracking and overseeing all changes. The CAB is focused primarily on reviewing change requests. The review should address risk associated with implementing and not implementing the change, including potential unintended consequences (e.g. the impact and relationship of this change on other changes).

The CAB fulfils its role by:

- Reviewing RFCs (seeking additional advice of subject matter experts, as needed)

- Asking probing questions to fully understand the proposed change and the business purpose

- Questioning the expected return on the investment required for the change

- Evaluating the proposed change for risks and determining whether the implementation plans appropriately address those risks

- Evaluating whether the proposed change is likely to produce the intended outcomes without adverse impact on the business

- Ensuring the change is evaluated in the context of any related changes

- Scheduling and prioritizing changes

- Ensuring the proposed timing is appropriate (doesn't conflict with business activities, other changes or operational activities)

- Determining the likelihood of unintended impacts

- Making recommendations to reduce risk, increase likely success and minimize adverse impact on the business

- Requesting a more in-depth, formal change evaluation for a given change if deemed necessary. The CAB uses the findings of the change evaluation to assess the change

- Ensuring there is clarity with respect to who will have the responsibility to implement, test and deploy the change

- Advising the change manager of its findings and recommendations.

The CAB can perform its role informally through email, or formally through a chaired meeting, where minutes are taken and people raise their hands to speak to the board. The most common form is an open dialogue, facilitated by the change manager. The CAB members discuss the various aspects of the proposed change. The change requester is present to answer any questions the CAB may have.

The culture and the business needs of the organization determine what's best in each situation, but I strongly recommend against a format that intimidates requesters and emphasizes the authority of the CAB. This format will become a significant source of cultural resistance and a liability to further maturity of change management.

3.2.6.2 CAB membership

The CAB is made up of two types of members:

- Regular (permanent) members

- Ad hoc members.

Permanent members are typically senior representatives from the various disciplines, each with broad and deep knowledge of IT generally, and their representative discipline specifically. These members are generally at all CAB meetings and provide ongoing expertise and continuity, which is important as the CAB's culture develops. In the first phase, it's not uncommon to have a broader permanent CAB membership. As it matures, there will be more reliance on a few core members and more expertise brought in only as needed, depending on the change under consideration.

Ad hoc members are chosen for their specific knowledge in a particular area and are called upon based on the nature of the request(s) under review. It is the responsibility of the change manager to ensure that the right expertise is available so that each RFC is adequately reviewed.

In this first phase, it's especially important to select the right people for the CAB. The credibility of the newly formed CAB will be largely contingent upon the combined credibility of its initial members. Careful attention should be given to select members who:

● Have strong expertise in one or more technical disciplines

● Have excellent communication skills

● Are widely respected in the organization

● Work well across organizational boundaries.

The CAB is comprised of technical staff and key decision makers. There are no set rules about who is on the CAB. The key is that each member possesses respected expertise and knowledge.

The CAB should include representatives from all IT functional areas and technical disciplines, key decision makers and business stakeholders, as appropriate.

Typical CAB membership:

● Senior network engineer

● Senior application developer

● IT operations managers

● Service desk representative

● Infrastructure engineers

● Senior security engineer

● Information security officer

● Standard change workgroup leads

● Business relationship manager(s)

● Service owners

● RFC author.

The change manager must analyse the RFCs scheduled for review at the next meeting and ensure the necessary specialized knowledge is represented at the CAB meeting. For instance, input from a senior network engineer would be of benefit when reviewing an RFC for changing network routing architectures.

3.2.6.3 CAB scope

One question often asked is, 'What changes should be submitted to the CAB?' The simple answer is that all changes to the production environment, or those that have the potential to disrupt production services, should be under change control. But that answer carefully dodges the real question, which is, 'Must all changes submitted to the CAB be under change control?' – and the simple answer is yes. In this first phase, we are introducing the concept of change control, and it would severely undermine that objective if we were to do otherwise. (I introduce standard changes and other means of optimizing change management in Chapter 5.)

Release management is covered in Chapter 4. You'll find a discussion regarding grouping changes into logical collections called release packages that are reviewed as a whole. I'll say more about these and other strategies for optimizing the CAB in Chapter 5. For now, suffice it to say that all changes are under the authority of change management. As the process matures, it works through multiple mechanisms – not just the CAB. Over time the CAB should become less focused on individual changes and more on the overall change capability, with the goal of improving change outcomes without direct review (again, see Chapter 5 on optimizing change management).

3.2.6.4 CAB authority

The change policy should be clear about change authorities. As stated before, the CAB is part of the overall change process but is not a decision-making body. The change manager is ultimately accountable for ensuring changes are approved/rejected by the appropriate change authority, and must take into account the complex and often conflicting interests of stakeholders – including the business itself.

The change manager will generally follow the advice of the CAB, but can approve changes (for many reasons) that the CAB has recommended against (and vice versa).

3.2.6.5 How the CAB works

The change manager compiles a list of RFCs that are ready for the next CAB review process.

Policy around submissions to the CAB should include a cut-off deadline, so the change manager and the CAB have time to review changes before the CAB review. Requests that come in after the deadline are generally pushed to the following meeting unless the RFC has some high business urgency or impact and the change manager decides to accept it into the current listing. Once the CAB list of RFCs is complete, the change manager notifies CAB members – often via an email that includes either a list of the RFCs or a link to a location where they can be found.

In the first phase, face-to-face meetings are preferred. If you have people at multiple locations, remote members can be included in the proceedings by using either voice or video conferencing.

The CAB meeting

CAB members are expected to review the RFCs prior to the meeting. They are encouraged to research the change, glean additional information from the requesters and consult with others to make sure they fully understand the change and its implications.

One of the common complaints of the CAB meeting occurs as a result of members who don't adequately prepare and instead waste time during the meeting by reviewing the RFC for the first time. Not only does this waste precious time, it is also disrespectful to the other CAB members who arrive at the meeting fully briefed. This is the prime reason for the cut-off deadline mentioned above – members must be allowed time (and are expected) to properly review change requests prior to CAB meetings.

There are no hard and fast rules when it comes to CAB meeting frequency. Though the most common frequency is weekly, many organizations have several shorter meetings during the week – often tied to release windows.

Other organizations have CAB meetings that are focused on a particular area, such as infrastructure, network or applications. In some cases, it may be desirable to have separate meetings with the business units affected.

The needs of the organization must weigh heavily on the frequency of CAB meetings. Generally speaking, less often can delay needed implementations, while more frequently can consume an inordinate amount of staff time.

Obviously, you should do what makes the most sense for your organization; be aware that how staff respond to the frequency will affect their support for the new change programme. I generally recommend weekly as a good starting point; be flexible and adapt to how the business and IT staff react.

The CAB agenda should include:

- Review new RFCs
- Review changes implemented since the last CAB
- Review proposed standard changes
- Review change schedule
- Consider continual service improvement (CSI) opportunities.

Review RFCs

This is the heart and soul of the CAB meeting – presenting and initial review of proposed changes. Make sure the change requester, key decision makers, and any technical experts representing the change are part of the meeting to answer questions about the proposed change.

Review implemented changes

The CAB should discuss the outcome of each approved change. For any failed or rolled-back changes, the CAB will need an explanation regarding why the implementation wasn't as expected and determine whether specific follow-up actions are required. Each failed change should be viewed as an improvement opportunity to learn more about the organizational change capability.

Many organizations hold post-implementation reviews (PIR), especially for significant or highly visible changes. This is an excellent practice and should ideally be carried out as part of the project management process, with the results being presented to the CAB to enable improvement opportunities to be assessed.

A PIR is much like a lessons-learned meeting. Its primary purpose is to review the change implementation in depth to determine whether the business outcomes were achieved. PIRs are most often conducted with full representation from IT customers and all key stakeholders.

The PIR should also identify issues or shortcomings during the planning and implementation, and make recommendations for improvements to the process or capability.

It is not recommended that the PIR is only undertaken as a follow up to changes that didn't go well. Doing so diminishes the value of learning from both successful and unsuccessful changes; it also casts the PIR as a punitive practice applied only to problematic changes.

Review standard changes

Standard changes are an important part of a healthy change management programme. I highly recommend implementing standard changes as early as possible in your change management implementation (see Chapter 5).

If you're using standard changes, proposed new ones should be reviewed at a CAB meeting in much the same way as normal RFCs are reviewed.

If operational issues relating to an approved standard change arise, the CAB should schedule a review to determine whether the approved process for standard changes requires modification.

In the normal course of CAB meetings, members should be mindful of any normal requests that might be good candidates to be considered standard changes.

Review change schedule

The change schedule is simply a listing of approved changes in a calendar format, and it should be consulted routinely when considering individual changes. It is important to review the overall schedule for any potential conflicts, incompatibility issues and improper sequencing of complex changes.

It is also necessary to ensure major changes are not implemented at times when the organization might be under strain. Be sure to compare change schedules with vacation calendars, staff training course dates or other activities that may impact on resource availability. The organization's schedules and cycles must also be considered. The CAB should be on the lookout for month, quarter and year-end times, school holidays etc., and also be aware of non-cyclical activities, such as trade shows, conferences, special sales promotions and product launches.

Consider CSI opportunities

Continual improvement is key to a healthy change management programme. This should be a core consideration throughout the process, and the CAB should be on the lookout for opportunities to improve both the change management process in itself and the way the CAB works.

This needn't be a complex or difficult task. A simple CSI register could be maintained listing ideas for improvements that get prioritized, assigned, and reviewed until implemented.

3.2.6.6 Emergency changes

A well-managed change process should match the rate of change required by the business. In other words, the bulk of changes should fall into the normal or standard category. If an emergency arises, however, it should be dealt with by an eCAB (a subset of the full board), as it might not be possible, or necessary, to gather together the full CAB within the emergency timeframe. The first question an eCAB must ask is, 'Should this RFC/change really be classified as an emergency or can it be deferred to the next regularly scheduled meeting of the full CAB (in which case, it's a high/urgent priority normal change)?'

The criteria for emergency change must be kept very high, and always for situations where there is significant and imminent risk to the business. There must always be provision for the special handling of emergency changes.

A closer look: urgent changes
Urgent changes should not be confused with emergency ones.

A very common criticism when adopting formal change management is that it's slow and bureaucratic rather than being responsive to the needs of the business. While the intent of change management is the exact opposite, this very issue can be a source of significant resistance and frustration, especially if the organization previously implemented changes ad hoc.

The argument is straightforward – whereas a change could previously be applied immediately to support the business, now a change request must be submitted, accepted and queued up for review at the next CAB meeting for approval. This is particularly problematic when the proposed changes resolve pressing issues or implement needed functionality. It may be tempting to use the emergency change provision to accommodate such change requests, but doing so severely undermines the spirit and intent of change management.

A well-established and communicated urgent change provision can take the wind out of these criticisms by providing a mechanism for accelerating the process to respond to unforeseen circumstances so that the business need is met. It is important to include the urgent change provision at the earliest stages of adoption so as to avoid unnecessary cultural resistance.

Urgent or emergency change?

The most common question around emergency changes is: how do you know whether the change you are being asked to approve relates to a real emergency? One of the challenges you face as a change manager is identifying real emergency changes from changes that were simply poorly planned or managed.

Far too often, the emergency change provision is used as a safety valve, of sorts. This misuse often stems from change management being viewed as primarily an IT internal process that stands in the way of doing things quickly. That is not at all the intention of emergency changes, and ramifications of treating it as such are far-reaching and should be avoided at all cost.

Emergency change status is reserved for those situations where there is a significant risk to the business – one that threatens life, financial loss or organizational credibility. A lesser standard opens the door for confusion and misuse.

Some organizations only allow emergency changes when there is a major incident underway and the incident manager is requesting emergency change to remedy a critical business-impacting issue.

This criterion pairs well with a process that treats emergency change requests with drop-everything urgency. It does not pair well with attempting to use the emergency process to compensate for lack of proper planning or simply ignoring the change process.

Obviously if there are emergency-level risks, timely resolution is critical. The concept is very similar to the business urgency and business impact chart associated with incident management. The response timeframe is determined by a combination of both urgency and impact.

In practice, change priority is a combination of business urgency and business impact of the proposed change, as shown in Figure 3.2. To be considered as an urgent priority, a change must be both high business urgency and high business impact, but this categorization does not make it an emergency.

Keep in mind that the vast majority of changes are in the planning and development stage long enough to give ample time to follow the normal change process. Urgent changes are those that are truly unexpected and require implementation at the next change window.

As I've said many times already, you have to do what makes sense for your organization, but systematic abuse of the emergency change provision will have a significant effect on the viability of the change programme.

Emergency change process

Given the above criteria, it is imperative that the emergency change process be extremely responsive and treats all emergency change requests with the highest level of attention.

Emergency change requests are most frequently submitted by email, phone or in person to the change manager.

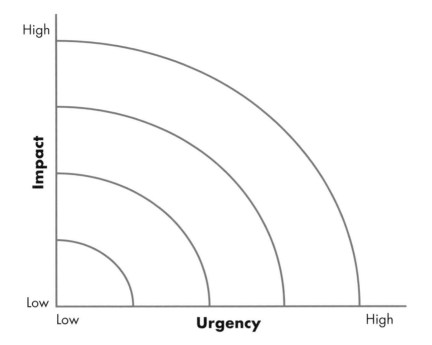

Figure 3.2 Change priority

The basic emergency change process is as follows:

- The emergency RFC is submitted to the change manager.

- The change manager immediately considers the request.

- If deemed an emergency, a CAB session (sometimes known as an eCAB) is called immediately.

- The eCAB can meet in person, via phone conference, email or other electronic collaboration method.

- The eCAB reviews the emergency RFC for the same concerns discussed in this chapter for normal changes.

- The eCAB makes an immediate recommendation based on the gravity of the situation and analysis of the risks.

- The change manager approves or denies the change and communicates the decision.

- The approved change is then authorized for immediate implementation.

- As a follow up, the emergency change is scheduled to be reviewed at the next normal CAB meeting (as with any implemented change).

The timeframes must be in alignment with the needs of the organization (government, defence, healthcare, banking and manufacturing all have unique challenges and drivers). Regardless of the business climate, it is best to adopt an emergency change process that more closely resembles a call to the emergency services than the normal change request process. This is for two very good reasons:

- Business outcomes are more important than processes, and a true business emergency demands an IT organization that can respond with changes at the needed pace.

- This level of responsiveness tends to discourage less critical changes from rising to the emergency level.

If your organization has a mature incident management capability – especially with provision for major incidents – you may want to consider delegating emergency change authority to the incident manager when a major incident is declared.

When a true emergency is being managed in this way, the incident manager is already entrusted with significant responsibility on behalf of the company. Because of the severity of the impact on the enterprise, senior business and IT management are involved. By design, decisions are elevated through the incident manager, who is responsible for seeking decisions from the appropriate authority and ensuring timely execution.

With that in place, then, it makes sense that the incident manager and the change manager work in tandem to secure and execute decisions from informed authorities in the organization.

I would only recommend this model for emergency changes in cases where there's an existing mature incident management capability.

As the change management capability matures, the number and percentage of emergency changes should be tracked as KPIs and kept to a minimum (exact percentage to be determined by the organization).

eCAB

Emergency changes are reviewed on an accelerated timeframe by the eCAB.

The eCAB can be an emergency call of the normal CAB membership, or it can be a completely different group. Either way, the personnel must be available at a moment's notice. Members are expected to leave meetings and drop normal workday tasks when it is convened.

In addition to the ability to assess technical risks and issues, eCAB members must also be able to assess business risk and impact. The eCAB is often heavily staffed with senior leaders and managers who are responsible to the organization for the management of business risk.

The change manager ensures the right senior management are on hand to address business risk and weigh in on the proposed change(s). This is important to ensure the actions of the emergency change are in alignment with the best interests of the company.

The eCAB must not allow the urgency of the situation to cloud their judgement in evaluating the proposed change for risks and unintended consequences, and they must determine whether the proposed change is likely to achieve the required outcome.

3.3 Metrics/KPIs

To ensure change management achieves the desired results, clear metrics must be established and regularly reported and reviewed. These operational metrics are intended to highlight the need for corrective action or directional changes in the change process.

Before metrics can be established, we must identify some higher-order objectives that the change management programme is seeking to deliver. These objectives form the CSFs and usually flow directly from the change policy's explicit objectives.

By nature, there should be a limited number of CSFs – generally two or three (and not more than four). 'Critical' in this case means those few core factors that determine success; they are generally objectives specified by senior management. The term 'critical' is used to indicate that CSFs are not broad, generalized statements of desired results; they are those determined to be vital to the organization and are, therefore, more important than other measures that may also be desirable.

CSFs are directional and can be quantified. They are not metrics themselves, but rather statements of desired direction, often involving the reduction or increase of a certain element, or an improvement of something in a stated direction. It is important to establish CSFs that are specifically suited to your organization, that support the current requirements of the business.

KPIs, by contrast, are selected to measure how well the process is supporting the desired outcomes (CSFs). In other words, KPIs are chosen to measure whether a CSF is being achieved (or not).

What is a KPI?

Key performance indicators (KPIs) are derived from established high-level critical success factors (CSFs) – those few core elements that must be in place in order to be able to judge whether a process is successful.

A small set of KPIs are established for each CSF. The operative word is 'key' – indicating that this particular measure has been determined to be a key indicator in support of the CSF. KPIs are quantitative and measure elements that indicate whether the process is moving closer to the CSFs or further away.

Together, CSFs and KPIs establish directional objectives and quantitative measures to demonstrate process performance in that direction.

3.3.1 Sample metrics for phase 1

Let's examine some reasonable objectives for phase 1. Keep in mind that you want to measure what's important in your organization at this particular point in time. Both CSFs and KPIs will change with time, experience and maturity.

In the first phase, it is important to establish objectives that work as CSFs. Under each of these, I've added some suggested KPIs that help quantify achievement of that objective:

- **CSF 1** Increase changes managed by new change process

  - KPI 1 Increase in changes reviewed by the CAB (percentage)

  - KPI 2 Reduction in unmanaged changes (percentage)

  - KPI 3 Increase IT staff satisfaction with change process (favourable survey results)

- **CSF 2** Reduce negative business impacts relating to changes

  - KPI 1 Reduced number and percentage of change-related incidents

  - KPI 2 Reduced business impact of change-related incidents (number and duration)

- **CSF 3** Ensure cultural adoption

  - KPI 1 Increase in changes that provide proper documentation for support staff (number and percentage)

  - KPI 2 Decrease in the number of unauthorized changes (those that bypass the change process)

  - KPI 3 Increase in satisfaction of change management for IT staff and end users (favourable survey results)

- **CSF 4** Manage IT environment without bureaucracy

  - KPI 1 Increase in changes completed in time and with estimated resources (number and percentage)

  - KPI 2 Increase in changes that delivered expected business outcomes (number and percentage)

  - KPI 3 Increase in customer satisfaction for change management from the project management office (PMO) and end customers (favourable survey results)

  - KPI 4 Reduction in number and percentage of urgent and emergency changes.

Note that KPIs can support multiple CSFs. Target goals should be established for each of the key areas. In this initial phase, I would highly recommend target goals that increase over time. For instance, you may establish a goal for changes managed by the CAB to be a 50% increase, followed by an additional 50% once achieved.

Stepping of targets is fundamental in a continual improvement approach – take small, achievable steps, and celebrate success, to build support and momentum for the programme.

3.3.2 Reporting

Regular reporting of the CSFs and KPIs are important to demonstrate early results and maintain ongoing support for your new change management programme. It's helpful for cultural adoption in that it demonstrates in tangible numbers how well the programme is working.

If staff or customers feel the new change programme slows down project delivery or doesn't produce results – even if it's only a perception – it fuels resistance to the new programme.

Starting with the very first adoption efforts, I encourage monthly reporting to senior IT management and key business stakeholders. Early and frequent reporting demonstrates transparency and encourages engagement in improvement efforts. Reporting should be focused on progress achieved and challenges faced – and include details of corrective actions needed to stay in alignment with stated objectives.

Reporting should never be allowed to become a finger-pointing activity. Notice that I'm not suggesting sorting compliance by individual teams.

3.4 When to begin phase 2 (success criteria)

How do you know when you're ready to move on to phase 2? If you already have a 'phase 1' programme in place, how do you know you're ready to mature it?

Recall that our phase 1 objectives had to do with successful programme adoption; briefly:

- Introduce change control
- Show early results, demonstrating value to the business
- Ensure cultural adoption
- Manage the IT environment while minimizing:
    - Bureaucracy
    - Unnecessary delays
    - Organizational resistance
- Set the stage for future maturity.

3.4.1 Metrics stabilization

The first sign that you may be ready to think about phase 2 is when you are consistently meeting the targets for metrics. Each company has its own unique culture, so it's hard to know exactly when that's been achieved, but in general, you're looking at those KPIs specified in section 3.3.1 and seeing target numbers and percentages achieved or even exceeded, plus positive feedback from the various stakeholders in the change management process.

3.4.2 Cultural acceptance

There is more to success than just metrics performance, however; you need to determine the state of the culture – which is hard to do with pure metrics. Regular surveys of IT staff and customers are important to gauge satisfaction with and engagement in the change programme. If staff are showing signs that they're getting tired of the new change programme, and there's a growing resentment, you're probably not ready to proceed to phase 2.

In the early stages, increasing the number of changes that go through the new change programme is a key factor. It's not unlike a parent holding a child's hand as they're learning to walk. Gradually, over time, the parent holds on less and less, eventually leading to that moment when the child walks with no parental assistance. Just like the child, your new change programme will have some falls and setbacks. Some of them may even be painful.

Signs that your change programme is starting to have legs include the enthusiasm with which changes are reviewed, and the collaborative spirit of the CAB members. Be on the lookout for staff who want to make improvements that increase the quality of the process (which is very different from suggesting that you stop doing change management!). When staff see the value of change management to the point they want to see it improve, you're well on your way to the next phase.

3.4.3 Business value

One of the key measures of phase 1 success is when the business recognizes the improvement in change execution and reduction in negative impact. The support generated because the business is noticing improved results from the change initiative does wonders for getting the support needed to move on to the next phase.

Show value before doing more
In general, it's always best to be successful in a small thing before moving to a bigger thing. Change management is no different. This is why it is so important to show early results on the investment you're making in change management. If your results are limited or marginal, or worse – poor results, or a reputation for slowing down delivery – you're going to have an uphill battle getting support for further improvements in the dubious change programme.

Value is always in the eyes of the beholder, which in this case is the organization. If only IT is convinced of the value of change management, it can be argued that it has no tangible value. Harsh as that sounds, it is always the customer who determines the value of what they receive in the way of service.

3.5 Chapter 3: key concepts

The key concepts in Chapter 3 can be summarized as follows:

- Start with a basic 'phase 1' process to ensure successful adoption.

- Introduce the concept of change control.

- Establish a weekly CAB review and select the right CAB members.

- Manage and monitor cultural acceptance.

- Measure adoption milestones and progress.

- Plan to show early results against organization needs.

- Create a change policy that clearly describes change management expectations.

- Establish a procedure for handling emergency changes.

- Move to phase 2 (see Chapter 4) when the organization is ready.

4 Maturing change management (phase 2)

The challenge we face in maturing change management is how to do more with less (rather than doing more by adding unnecessary complexity). The essence of an effective change management programme is achieving the objectives while keeping complexity to an absolute minimum.

Where complexity is required, it must be thoroughly weighed against the business value it creates, while considering alternative means of achieving the same outcome(s) in a less convoluted way.

Simplicity is the operative word and must be kept at the forefront of change management adoption.

The basic change management programme described in the previous chapter must be handled carefully to avoid the pitfalls that so frequently accompany too much focus on process and too little attention to the business outcomes.

In phase 2, we'll focus on optimizing business value by:

● Proactively managing the change lifecycle (including integration with project management and IT governance)

● Shifting the focus onto the achievement of business outcomes

● Optimizing the effectiveness and efficiency of the change process.

4.1 Maturing a basic change management programme

This section applies as much to a newly adopted basic change management programme (as described in Chapter 3) as it does to an existing change management effort. The aim is to build a change capability that's constantly improving to meet ever-changing business needs. Many organizations, however, already have a change programme similar to the basic one and, too frequently, even a programme that has not stagnated becomes bureaucratic and ineffective over time. When bureaucracy starts settling in, cultural dissatisfaction and resistance are not far behind, and you risk losing the momentum your change programme once had.

In this chapter, I'll describe how to maintain or regain momentum from phase 1, and increase the effectiveness and efficiency of your change capability.

4.1.1 Governance compliance

Corporations and public agencies operate under regulatory and governance compliance requirements. Though specific requirements vary depending on the country, size and nature of the enterprise, the ability to demonstrate effective control over the IT systems that underpin the organizational business is important.

The Sarbanes-Oxley Act and Health Insurance Portability and Accountability Act (HIPAA) are two US examples of laws with strong control requirements. Other countries and industries have others and more are being enacted regularly.

The degree to which your organization must demonstrate regulatory compliance must be understood. A central theme emerges: organizations must be able to demonstrate management of IT changes that include current (up-to-date) process documentation and logs of process compliance audits.

This single factor requires a formal lifecycle approach to managing changes. A last-check, IT-centric change programme is not sufficient to demonstrate effective change control and may not be fully compliant with regulations or laws.

Regardless of the specific requirements to which your organization may be subject, effective change management requires that changes are:

- Documented

- Evaluated for risk

- Prioritized and authorized by management

- Able to be escalated as emergency changes

- Tracked and reported.

These should be accomplished as efficiently as possible, of course, but compliance must be auditable – meaning, there must be documented evidence of the activities.

4.1.2 Change in scope and configuration management

In phase 1, the scope of change management is somewhat vague but generally understood to be any changes that could have an impact on the IT services used by customers. Because we focused on the introduction of the concept of change control, it was more important to start with the addition of formal management of changes.

As change management matures, it will become more important to have a clear and consistent understanding of the increasing scope.

Change management is closely tied to configuration management. Configuration management supports change management by providing a logical view of the infrastructure. It allows change management to answer a question such as: What configuration items (CIs) are impacted by this change? Understanding all the components of an IT service, their current state and how they are related to each other is a powerful enabler of effective change evaluation. Without configuration management, change management must rely solely on design documentation and the memories of the IT staff who built the system. Both of these have practical limits.

On the one hand, with effective configuration management the scope of change management is quite clear – if it's in the configuration management system (CMS), all changes must be managed by change management. On the other hand, without configuration management in place it's a bit more challenging, but the goal is the same – if it could impact on a production IT service, then it should be under change control.

Obviously, configuration management is beyond the scope of this publication, but it's worth noting that it's difficult to move beyond the basic change programme described in Chapter 3 without effective configuration management. (It's also worth noting that change management must update the CMS to reflect approved changes.) Obviously this requires change management to know which components (CIs) are associated with any requested change and how, specifically, they are being changed.

The short story is that all IT services that support business processes and all the components that underpin those services must be documented (in the CMS), and changes to them must be managed through change management.

While this may sound a bit academic, it is an important distinction that links to the reason why we manage changes in the first place – to support business outcomes. That being the case, any changes to infrastructure components that are associated with business-supporting systems must be managed to ensure that the required levels of security, availability and integrity are maintained. Analysis of how any given change may impact on other parts of

---

Are changes in development and test environments subject to change management?
A common question around change management is, 'Do changes in development and testing have to go through change management?'

The question demands further clarification. The development and testing environments are critical to the effective validation of changes supporting business functions. So, yes, changes to those environments themselves are clearly under change control. But if the question is, 'Does change management need to review and approve changes (being developed) moving through the development and testing environments?', then the answer is no.

The entire development process is itself a business process, making the infrastructure upon which the development process is conducted a production environment.

Further, development and testing must be rigidly maintained such that they accurately reflect the production environment, otherwise testing conducted there is not an effective validation of performance to be expected in production, which increases the risk to the business of changes.

the infrastructure and/or IT services is an essential element afforded by effective configuration management.

Either way, the change policy should make the scope for change management clear. Note: as the change management matures, the scope of the process may be updated, too. As the scope changes from the basics discussed in the previous chapter, it is critical that the updates are communicated.

4.2 Phase 2 goals

You'll recall our goals in phase 1 were:

- Introduce change control
- Show early results
- Ensure cultural adoption
- Manage the IT environment
- Set the stage for future maturity.

These are carefully chosen to maximize the likelihood of a successful platform upon which you can mature the change management capability.

With these secured, we'll be adding some additional elements that will maximize both business value and effectiveness in managing IT changes.

Phase 2 goals are:

- Manage the change lifecycle by:
  - Effective and efficient management of changes
  - Demonstrating appropriate governance
- Maximize business outcomes
- Optimize change management.

4.2.1 Manage the change lifecycle

One of the significant limitations of a phase 1 change process is that it is entirely reactive. The only task is one of inspection after the planning and development work has been completed and before the change is deployed. This last-check-only change process causes staff to view change management as largely ceremonial, focusing exclusively on the communication of changes that are already happening – as opposed to a control process whose function is to actively manage all changes.

Although there is some value in this approach, it is limited by its exclusive reliance on direct inspection of all changes. In other words, change management has no ability to affect the quality of changes outside of reviewing those brought to the change advisory board (CAB) after the development work is complete. It also has no ability to reject changes that might be wasteful, risky or unnecessary before any work is started.

This is an important point for maturing change management – breaking the paradigm of the CAB (and with it change management) from a reactive-only review board to one where requests for change (RFCs) are evaluated before any work is authorized or started.

4.2.2 Maximize business outcomes

Managing the change lifecycle and optimizing the change process are not only the logical next steps in maturity, they are integral to achieving it. This includes consideration of the business outcomes that a change is intended to support, not after the implementation but before it is authorized to proceed into development. This change-approval/work-authorization responsibility is an entirely new aspect of change management, although one in which many IT organizations never engage. I believe this omission is the single greatest reason why change management is widely viewed as a bureaucratic process that slows down delivery.

By contrast, I contend that achieving business outcomes is the core of effective change management. The goal must be for change management to be an organizational (not simply IT) capability that ensures that changes produce the results the business requires. Anything less is a misuse of organizational resources and should be challenged by organizational governance (senior management, boards of directors etc.).

This means the RFC and those reviewing it must address not just the technical and logistical aspects of a change but also the expected business results – the outcomes that the business is hoping to achieve.

4.2.3 Optimize change management

'Optimize' can mean many things to many people, so perhaps it's best to start with what I mean by the term. I see it as finding the right combination of operational practices to produce peak business value, at the same time keeping costs in the form of labour and delayed realization of value to a minimum, and with no unnecessary process bureaucracy.

The next chapter is dedicated to optimizing change management. At this point, suffice it to say that what we're trying to achieve with the addition of the pre-development checkpoint is to strike a better balance of process and outcome, while managing business risk.

I cannot emphasize enough that adding this checkpoint must be carefully managed so it does not become a no-value step that just slows down progress and causes resentment.

The tale of bad water

The old story goes that a certain town had great-tasting drinking water that came directly from a clean river that originated from the nearby snow-capped mountains. The town took great pride in its water.

One day, however, the water started to taste bad. Rumours went around that it might even be dangerous to drink. The town's water treatment plant was fitted with expensive filtration equipment, but the water remained undrinkable.

Consultants were then hired to analyse the problem. They concluded that the treatment plant required significant upgrades to restore the town's treasured water quality. The town council considered their options.

One day, a young man out walking saw a dead moose in the middle of the river. Not long after he'd returned home and told his parents about it, a group of townspeople went out and, with much effort, removed the dead animal from the river.

Not long after that the town's water was crystal clear and tasted great once again.

The moral of the story?

Sometimes you need to go upstream and remove the dead moose from the river.

4.3 Phase 2 process

In phase 2, we are moving upstream, if you will, adding an extra checkpoint before development work actually begins. This pre-work review allows change management to expand its focus on the end-to-end change lifecycle. Figure 4.1 shows the simplified view of the newly added step.

This new checkpoint is achieved with the same CAB described in the previous chapter but with some key differences. The most notable is that change management becomes the actual authorizing agent for development work to begin – nothing can start until an RFC has been approved.

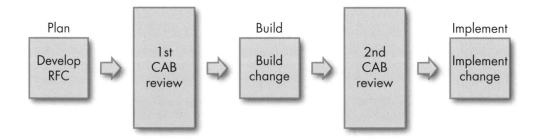

Figure 4.1 Expanded basic change process flow

In organizations with a strong project/portfolio management capability, this function may already be performed to some degree by the project management office (PMO). If you're fortunate enough to have effective project governance, this additional CAB function should be integrated with the work the PMO is already doing. In this case, the PMO may give the approval for the project, taking into account financial analysis (return on investment (ROI), net present value (NPV) etc.) as well as business prioritization and portfolio optimization.

If your organization follows the systems development lifecycle (SDLC) methodology, you may recognize the phases. In phase 2, we're inserting a CAB checkpoint between the SDLC design and build phases, as shown in Figure 4.2 (shown also is the previous post-development CAB between testing and deployment SDLC phases).

As you look at inserting this new CAB point, be mindful of how it intersects with other processes, especially IT governance and project management.

RFCs are reviewed for technical and logistical details to ensure the proposed change:

● Is likely to achieve the stated business outcomes

● Adequately addresses risks

● Is unlikely to have an adverse impact when released.

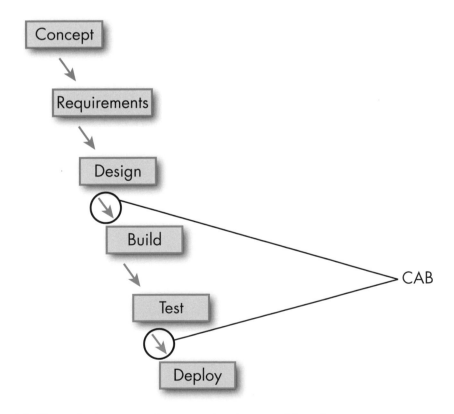

Figure 4.2 Phase 2 change and systems development lifecycle

Expanding the CAB role

Unfortunately, reactive-only change management is so prevalent that many IT people work their whole career without ever imagining that change management has any role other than the final check before going live. For them, it feels like an enormous overreach where change management exceeds its rightful domain and meddles in the planning and development process.

You're asking people to literally rethink everything they've known about IT delivery and change management, and it's important to not underestimate the scope of the change this new role presents.

People with a 'traditional' view of change management are much more likely to resist process additions that they think are irrelevant, produce no value, disrespect staff expertise and add bureaucratic bloat. In their minds, such additions are contrary to delivering business results, so introducing change in a heavy-handed way (especially when staff believe the current process has been serving their needs quite adequately) could provoke awkward debates and prove unproductive.

This is why it is important to focus on outcomes over process. Start your change maturity programme by communicating the goals, which now include maximizing business outcomes. This is a goal with which it's hard to disagree. Starting from a point of agreement on goals is important for getting the IT culture to buy into these process additions.

As with all good communication, you want to start with a solid reason why you're making changes to the change management process. Considering why, and what's in it for them, is fundamental to gaining cultural support for changes.

This new checkpoint greatly expands the role of change management beyond its traditional 'do-no-harm' final check.

Many change management programmes become stymied at this very point, and the longer you remain at the phase 1 level, the harder it is to move into phase 2 maturity.

Crossing over can be a challenge that has more to do with people and organizational culture than with process.

The keys to success are:

● Careful transition planning

● Gaining an understanding of and enabling integration with existing planning and governance

● Clear and consistent communication of what's changing and why

● Clarity in (especially new) roles and responsibilities.

Where there's significant cultural resistance, or lack of management support, organizations tend to compensate for immature change management by offloading this critical review process to other areas, typically development or architecture teams.

The upstream roles must work very closely with the rest of change management for the entire process to be effective. The task of consolidating the change-related functions performed in other locations into a more central change management capability is likely to produce more initial resistance than value.

Keep in mind that this second phase expands the focus to include business outcomes versus just the test and review that was part of phase 1, but don't attempt to turn it into a monolithic change management process. If the functions being performed elsewhere meet all or part of management of change lifecycle, keep them.

Recall W. Edwards Deming's insistence that quality cannot be inspected into the process; it must be part of the overall change management flow. The sooner an error is detected and corrected, the lower the cost and risk to the organization. By introducing this next stage of maturity, change managers are in a position not only to review and validate proposed changes before development work begins, but also to be involved throughout the lifecycle of the change.

Change lifecycle

All changes have a start and end point. In IT service management (ITSM), we refer to the end-to-end process as the lifecycle of a change (or a service). The application of an end-to-end approach is one of the ways to improve delivery of business value The lifecycle starts the moment an idea for a new or changed service is communicated and carries on through the entire effort to bring that change to fruition.

Although the change is theoretically completed once it is implemented and producing the value the business needs, in reality there is no real end. Services in production must be continually improved to meet ever-increasing business needs. Over time, previously released changes peak and fade, new changes emerge, and the lifecycle continues. It can be thought of more as a continuous flow than as a static point in time.

Effective change management must accommodate changes of all sizes and types and that come from multiple sources, such as:

- Continual service improvement (CSI) efforts
- Incident-driven changes to correct issues
- Business-driven changes to support business outcomes, such as:
  - Bringing in new IT services
  - Adding functionality to existing IT services
- Maintenance and upkeep.

Each of these requires different levels of change management oversight, but all must be effectively managed.

The view of changes from a lifecycle perspective is a departure from the traditional understanding of change management and is one of the key points that must be carefully communicated for a successful change programme.

Once an RFC has been raised, and initial requirements have been gathered, it's time to draft a proposed solution to meet the requirements. The plan should be documented in enough detail that it can be understood and reviewed by technical staff and IT decision makers. In this context an RFC can be viewed as a business case for the requested change. As such, it becomes a decision support tool that should include information about rationale, costs (which may include development and operational cost estimates, NPV and/or ROI analysis options). Of course, as in phase 1, the RFC includes all regular information, including expected business outcomes.

The CAB reviews the RFC. If and when it is approved, the development team(s) can be authorized to begin work.

When development and testing are complete, the RFC is brought back to the CAB, where it is reviewed for potential production release. This review ensures that the required results described in the proposal have been effectively met by the solution developed.

If approved for release, the change is then implemented in production and enters a stabilization period. When the acceptance criteria are met, and the business has accepted the change, the CAB does a final review, at which time the change is declared complete (and successful), and the RFC is closed out.

4.3.1 Process flow

In Figure 4.3 we take a more detailed look at the change lifecycle. We're adding a single new process step – engaging the change management process from the very beginning of the change lifecycle. Regardless of how formal, every change has some form of requirements gathering and planning – this is basic to project management and required for a successful development process.

All the information gathered in the early requirements and design phases is documented in a change proposal that is presented to the CAB for review before work begins.

Change proposal

All significant changes should be documented in an RFC that is submitted to change management. The RFC must include a high-level description of the proposed change, which might be a new service, a significant change to an existing service, or decommissioning of a service. Especially important is the inclusion of the desired business outcomes, which should be described in sufficient detail that their achievement can be demonstrated through quantitative measures.

Depending on the organization, RFCs may include a business case that addresses the relevant business issues, risks, and considered alternatives (see Appendix C).

The proposal should also describe a high-level schedule so change management can assess the impact of this change on others in development.

The proposal should include a functional block diagram of the proposed solution, noting new and existing components and how they must be adapted to support the new change.

Requests for smaller changes, such as server and infrastructure patching and operating system upgrades, generally include a before/after state diagram and description, along with a high-level plan of how the change will be implemented (phased in, all at once, regional roll out etc.).

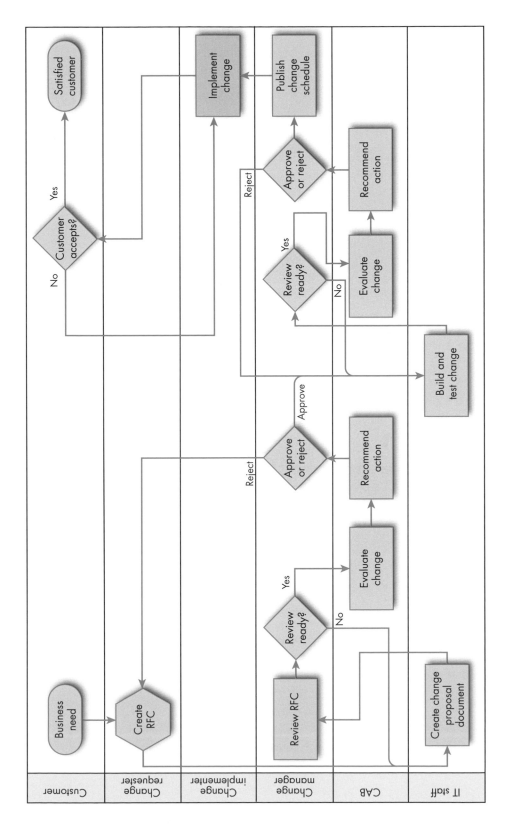

Figure 4.3 Phase 2 change management process flow

69

4.3.2 Phase 2 CAB

Recall that in phase 1, the CAB was primarily concerned with avoiding negative business impact by evaluating developed and tested changes for risks of unintended consequences and logistical challenges. Change management was not involved before the change had been developed and tested, so it's no wonder that there was very limited ability to avoid issues resulting directly from how the change was developed.

In phase 2, the RFC must first and foremost be explicit on the intended business outcomes. For change management to effectively evaluate the proposal's likelihood of producing the desired outcomes, they must be clear and unambiguous. To the degree possible, they should be stated in quantifiable terms such as:

- Results where an increase/decrease is measurable. These could include:
  - Transaction throughput rates
  - Server response times
  - Number of simultaneous customer interactions
- New functionality reflected in:
  - The number of new use cases
  - User interface feature requirements.

When reviewing the RFC, the goal is to think through the proposed solution and identify any potential issues with how the solution is designed. The change proposal must address any business or IT process changes to support the changed service and describe how the proposed deployment will work.

The review needs to be undertaken by well-qualified staff who look at the big picture.

Consider:

- Are the desired business outcomes documented, understood and approved (by the business)?
- What are the risks to the business and/or IT systems (risk of proposed change versus not making the change)?
- Is the proposed change the most effective means of achieving the desired business outcomes?
- What effect will the proposed change have on existing business and IT processes?
- What is the relationship between this change, other changes and existing services with respect to data integrity, performance, system availability and continuity?

- Can the infrastructure support the change (network bandwidth, firewall rules, server capacity etc.)? (Does the change propose additional infrastructure capacity, and is it sufficient?)

- Does it follow established technical and architectural standards?

- Does it introduce new technologies, and if so, are we prepared to support them?

- Does it add unnecessary complexity? (Is there a simpler solution?)

- Does the organization have the knowledge, experience and capacity to deliver the proposed solutions? (Does the proposal describe how these will be accomplished – training, partnerships, consulting etc.?)

- Does the proposal address the business process and cultural transitions required by the change?

- Is the business likely to realize the desired outcomes?

- What is the impact on service?

- What is the impact on customers?

- What is the priority for this change?
    - If everything is high priority then nothing really is
    - If everything is designated as high priority, this impacts on resource availability and risk. Change management should not approve the RFC without considering both these factors

- Are all stakeholders/constituencies in the change aware and sufficiently engaged to provide feedback?

Change management is intended to be a control, communication and coordination process, with the responsibility (and authority) to ensure that approved changes align with business needs and can produce the desired business outcomes without undue risk to the business. The business must be apprised of related risks and consulted on the acceptable level of risk (of making versus not making the change).

This new step provides a clear separation between change planning, change development and change implementation. In this way, change management is responsible for the control, coordination and communication of changes throughout their lifecycle. At any point in time, change management must be able to articulate the current state and progress of all changes under consideration.

Once the CAB has completed its initial review, the change manager will ensure that the appropriate authority either approves the change to move it into execution (build) or rejects it for further planning.

As in phase 1, following development and testing, the change returns to the CAB for a final review before authorization to release into production is given.

This post-development review can now focus on the following questions:

- Does the change achieve the stated business outcomes? (Does the testing verify it?)

- Has the build and test process identified any additional risks, and have they been addressed?

- Is the business ready for the change at the proposed time? (Has there been effective communication?)

- Is the implementation plan realistic and complete? (Are IT staff trained and prepared for a successful implementation?)

- Does the test plan appropriately validate the change success, the continued stability of the service and the acceptance criteria defined in the change proposal?

- Is the roll-back plan realistic and complete? For major releases, has the roll-back plan been tested and proved capable of restoring the pre-implementation state?

- Are there any conflicts or risks associated with the targeted change window?

4.3.2.1 CAB membership

In this second phase, we'll add some more CAB members:

- Business relationship management

- Service and process owners

- Business representatives (may vary with the RFCs under review).

With the focus on business outcomes, business representatives and relationship management are important to include in change review. Service and process owners are included because process changes are within the scope of change management, and process owners are ultimately accountable for how their process meet the needs of the organization.

These new members have the same collective responsibility to evaluate proposed changes using their knowledge, experience and expertise.

4.3.2.2 Formal change evaluation

In cases of major changes, or where complexity or risk is particularly high, the change manager may commission a formal change evaluation process. This evaluation can be performed by a team of specialists with the appropriate skills and background to conduct an in-depth analysis of the proposed change. The services of specialized consultants and representatives from key vendors, partners and component providers may also be engaged.

The end result is a formal report detailing the findings, identified risks, and risk analysis, and may include recommendations for risk mitigation or compensating measures to be included in the change development effort. The report is provided to the change manager and typically reviewed at a CAB meeting.

4.3.2.3 Emergency changes

Emergency changes are intended for circumstances where the company is facing dire circumstances – significant risk to the organization that requires immediate remediation. Emergency changes are one of those governance and compliance points where the process must be well-defined and documented and followed consistently in normal operations.

It is important that the intensity of the situation surrounding emergency changes does not derail the documented emergency change process. Although the rules of engagement are different when there's a major incident that requires an emergency change, the requirements for effectively managing changes remains the same.

With that said, a common approach in organizations that have strong incident management is for the incident manager to work closely with the change manager to facilitate timely review of emergency changes.

All emergency changes are still subject to tracking through the normal RFC tool, reviewed by the eCAB (or some other form of review), and are ultimately approved by the appropriate change authority. Urgency, and even dire situations where the business is at significant risk, do not justify violation of change authority.

Good practice is to leave such emergency changes open through a stabilization period. If the organization requires post (major) incident reviews, the results of that review would come to the CAB and be included in the CAB review of the implemented emergency change, ultimately leading to the normal change manager approving the closing of the emergency change.

These pieces can be accomplished in various ways, but keep several good practices in mind:

- Full delegated approval authority can be granted to a major incident manager.

- The scope of delegated authority is limited to changes directly related to remediation of the major incident.

- Full delegated approval is valid for the life of the major incident, and is withdrawn at the resolution of the incident.

- Emergency changes should be tracked using the same tracking method as all other changes, and with the same expectation of the level of detail. (Moments of crisis can present elevated opportunities for errors and oversights that can have far-reaching impacts on the organization. Such changes should be handled at the highest standards, not the lowest.)

- Emergency changes should be reviewed by the normal CAB process after the fact (this is especially important for authorization of CMS updates).

For organizations that adopt this form of emergency change delegation, there must be complete clarity in roles and responsibilities for both the change manager and the incident manager in these circumstances, and it should be spelled out in the change policy.

4.3.2.4 RFC status tracking

One additional clarification on the role of change management – the change manager (in conjunction with delegated approvers) approves the updating of the status of changes as they progress through the workflow. This is important because, first of all, change management is responsible for managing the lifecycle of changes. Secondly, changes must meet certain criteria before their status is updated. Change requesters and implementers communicate the progress and status of changes in implementation, and the change manager must determine whether the criteria for a change of status have been met.

The results of an implemented change will generally fall under one of the following statuses:

- **Successful/implemented** Successful changes are pretty straightforward. The change was implemented with minimal deviation from the documented implementation plan and has met the acceptance and stabilization criteria.

- **Rolled back** Changes can be rolled back for numerous reasons, including that important issues arose during implementation requiring significant rework of either the implementation plan or the change itself. Another reason would be that the implementation could not be completed during the approved change window and continuing with the work beyond that would disrupt the business.

  All RFCs must have a realistic, documented roll-back plan, which should state what decision-making authority the business is granting to the change implementer for roll-back determination. This ensures that all changes attempted will either be successful within the defined change window or rolled back, according to the documented plan and with no harm to the business.

  In many cases – especially with major changes – the business stakeholders may choose to be directly involved with the decision to roll back or continue in the face of issues. At their discretion, the business may approve less-than-optimal results in preference to rolling back (and forgoing the value the change brings).

  Rolled-back changes require change management and all change stakeholders to understand why they occurred and to ensure that the underlying issues are addressed before the change implementation is attempted again.

- **Failed** Not all organizations differentiate failed changes from those that are rolled back, and I see little reason to build an arbitrary distinction here; however, 'failed change' is a common phrase, so let me at least address it.

In my view, a failed change is simply one that hasn't gone as planned, or isn't producing the results expected. As I mentioned earlier, the business can choose to accept an underperforming change if it believes that the business value is greater than the negative impact. In such a case, the change could be classified as having failed, but the decision was made to not roll it back (at least not at this time).

Alternatively, a change could be so damaging that it must be rolled back; in which case, it's a failed change that's been rolled back.

The most important thing is that the organization has clear definitions for change outcome categories. Also keep in mind that, while it's beneficial to use such definitions consistently with the best-practice recommendations, it is more important that they serve the needs of your company.

Achieving business outcomes

In a manner of speaking, a last-check CAB (as described in the previous chapter) can be viewed as playing to 'not lose'. In other words, the best a reactive-only change programme can do is avoid obvious implementation problems and disasters.

Managing the lifecycle, on the other hand, is 'playing to win'. The whole point of managing IT services is to support the organization with services that maximize value while managing risk and minimizing negative impacts. Any change that falls short of achieving optimal business value is a poorly managed change.

There are a couple of implications here that may not be obvious. For change management to ensure business outcomes are achieved, the desired outcomes must be documented as part of the RFC. Further, business outcomes are determined by the business, and it is unwise for IT staff to make assumptions about what they may be generally, or in particular, for any given change.

Extortion changes

I've seen changes allowed to go into production with known issues simply because 'so much time has been invested' and 'the business needs it now'. Had the proposed changes been reviewed before building, many issues could have been avoided altogether.

I've also seen unplanned infrastructure changes made literally at the last minute, because an implemented change was incompatible with another component (i.e. the new application wouldn't run on the version of the operating system that was being used, necessitating an urgent upgrade).

I call these 'extortion changes', and they pose significant risk because they typically bypass normal planning, testing and review. They can also cause collateral damage to other services, causing a cascade of unplanned changes in production.

These types of change are particularly dangerous without configuration management in place, as the trickle-down effects to other services may not be identified until days or weeks later, when the initial change has been long forgotten. This makes it hard for support staff to identify the issue because the extortion change was made in haste and poorly documented.

This is not good practice in change management, though it may be common in some environments.

A second implication is that specific, non-ambiguous measures must be established to enable change management to determine whether a change has achieved the desired outcome(s). Change management must fully embrace this role and not default to the technical and logistical details only. Again, this is the essence of mature change management.

Stabilization period

After a change has been implemented, it should remain open in some fashion until the acceptance criteria have been reviewed and demonstrated to be complete. The stabilization timeframe should be defined in the change proposal, along with the acceptance criteria. For example, it's not uncommon to define target numbers of post-implementation incidents or business transactions to be processed. During this period (also known as early life support or warranty period), change management still has jurisdiction over the change. Therefore, should significant issues arise, it is the role of change management (in conjunction with the incident manager) to determine whether to roll back the change as per the roll-back plan in the RFC or see if additional changes are required that will successfully address the issue.

Once the stabilization timeframe and criteria have been met, with an acceptable number of incidents, the CAB will review the change, along with test results, incident trending reports and any other meaningful demonstration that the change has met the defined acceptance criteria.

What if the change can't meet the acceptance criteria?

In some cases, a change is implemented and the established acceptance criteria are never met, or the change may not produce the desired business outcomes. Does that mean the change was unsuccessful, rolled back, or failed? In these cases, there are several choices. Which is appropriate depends on the needs and desires of the business.

Accept 'as is'

The business can decide to accept the change as released, noting that it didn't meet certain criteria. If the issue causes service degradation or downtime, the business can decide if the downtime (and any associated workarounds) is acceptable, especially if the change has critical value that outweighs the negative impact.

This may be final, or the missed requirement can be prioritized for a future release. If the latter is chosen, the change should be considered successful, and the missed requirement will be carried forward in another change.

Roll back

If a criterion is not met that is significant to the business, it may choose to roll back the change until it can be reworked and released again with assurance that the missed requirement is met.

For organizations with a strong release capability, this may be little more than rejecting the current release, with the expectation that the next scheduled release will provide the required functionality.

Conditionally accept or reject

Lastly, the business may conditionally accept (or reject) the change and draw up a plan for remediation of noted issues. This may be in the form of a service improvement plan or simply as something to be addressed in the next release. A conditional acceptance constitutes a commitment of time and resources to address the issues. To maintain credibility and support, such commitments should be carefully considered, and always followed up.

Post-implementation review

Depending on the size and strategic nature of the change, a post-implementation review (PIR) may be conducted. This is standard practice in project management, and if a change is managed as a traditional project, this will be done under the project management office (PMO).

For smaller changes, this is carried out as part of the CAB review to close out the RFC. Not every change requires a full PIR, and it's a good idea to identify criteria in the change policy.

The culture of the organization will dictate the format and requirements of a PIR, but at a minimum, it should address the following questions:

- Were the stated and agreed outcome requirements met?

- Is the new/changed service operating as expected?

- Has the change been successfully documented and communicated to the support teams?

- Are there any open issues that could impact on the operational stability of the change? (How will they be resolved?)

- What went well?

- What could have gone better? (Are there specific opportunities for improvement that should be implemented in the organization?)

- Are stakeholders satisfied with the outcome of the change?

4.4 Metrics/KPIs

As before, critical success factors (CSFs) and key performance indicators (KPIs) should be carefully chosen to measure the specific needs of change management in the adopting organization.

Below are some examples of metrics that focus on measurable aspects of this second phase, and is in no way exhaustive.

- **CSF 1** Ensure efficient use of IT resources

    - KPI 1 Increase in changes completed within estimated time and resource usage (number and percentage)

    - KPI 2 Decrease in urgent and emergency changes (number and percentage)

- **CSF 2** Ensure effective management of change-related risk

    - KPI 1 Decrease in changes rolled back, failed, or requiring remediation

    - KPI 2 Decrease in unmanaged changes (number and percentage)

    - KPI 3 Decrease in differences between approved infrastructure state and CMS

- **CSF 3** Improve effectiveness of the change management capability
  - KPI 1 Increase in changes that deliver agreed outcomes (number and percentage)
  - KPI 2 Increased satisfaction for change management expressed by PMO and customers.

4.5 Chapter 4: key concepts

The key concepts in Chapter 4 can be summarized as follows:

- Change management is engaged at the beginning of the change lifecycle.
- RFCs must be authorized prior to beginning development.
- Completed and tested changes are brought back to the CAB for post-development review before release.
- Metrics that monitor efficiency and effectiveness are established.
- Clear and consistent reasons for changes are communicated to change management.

5 Optimizing change management (phase 3)

> *They always say time changes things, but you actually have to change them yourself.* Andy Warhol

Change management must be adapted to fit the needs of the adopting organization. In this chapter I'll cover a number of approaches to optimizing aspects of change management that help with its adoption.

5.1 Reducing changes going to the CAB

Depending on the size of your organization, the volume of changes to be reviewed can easily exceed the capacity of a weekly change advisory board (CAB) meeting. In large organizations, it's not uncommon to have dozens or even hundreds of changes every day. Obviously in these cases, a weekly CAB meeting is insufficient.

As change management matures, and more changes are brought into the process for review, the CAB meetings grow longer, making for marathon sessions that are unpopular, which hurts the cultural adoption process.

One approach is to classify changes into groups, so members and requesters can attend shorter, focused CAB sessions. These can be held on the same day, back to back, or additional weekly meetings can be introduced – infrastructure changes one day, application changes another, for example.

Another challenge a high volume of changes presents is that all changes tend to get lumped together, where large and potentially high-impact changes receive less review than is perhaps warranted, and smaller, more common changes get more than their risk deserves.

It's a pretty straightforward supply versus demand equation, where you can either increase the capacity of the CAB to process requests, or reduce the volume of changes requiring CAB review.

By far, the best approach is to reduce the amount of changes coming to the CAB, while still achieving the fundamental goals.

5.1.1 Standard changes

A frequent criticism made by change management and the CAB is that many changes are minor and rarely cause issues. It's both wasteful of time and disrespectful of staff to require these types of change to come to the CAB. Establishing standard changes allows for appropriate handling of these types of change, while maintaining appropriate oversight.

Let's start with a quick description of standard changes. Standard changes are:

- Carried out frequently in daily operation

- Considered low risk and non-complex

- Reviewed by the CAB and approved by the appropriate change authority

- Approved for a specified period (e.g. length of a volume purchase agreement) and regularly reviewed to ensure the change type 'standard' still applies.

Standard changes are a form of delegated change authority (see section 5.1.2) where identified (standard) changes can be executed at will at the local level. It's not that standard changes cease to be proper changes; it's that the authority to approve their use in daily operations has been delegated to a more local authority based on analysis, review and acceptance of the risks.

How to deal with daily operational changes
Changes that are performed so frequently that it's impractical for them to be reviewed by the CAB are a challenge in many organizations. On the one hand, by their very nature, these types of change cannot wait for the weekly CAB meeting to be reviewed, as that would hamstring any organization, and would rightfully be viewed as 'bureaucracy'.

On the other hand, ignoring them and hoping for the best isn't a good option either.

IT environments are very complex. Incidents are often the result of an intricate combination of changes or events. The most minor of changes, when combined with unknown other environmental factors, can cause problems. If operational changes are made without change management knowing, incident management will not have a complete view for effective troubleshooting, which will contribute to longer resolution times.

Even the most minor changes pose some risk, but the fact that many operational changes happen daily makes them good candidates for being classified as 'standard' and being placed under the authority of the operations manager.

Standard change is therefore an effective means of managing the risks of frequent operational changes without the increased overhead of bringing each instance to the CAB for review.

To facilitate their timely implementation, maintenance changes can be proposed as a standard change. For this to work effectively there must be clear criteria for what constitutes a maintenance change and responsibility for approving the list of such changes must be assigned. Maintenance changes are tracked through the reviewed/approved maintenance process, and can be accessed by support staff.

5.1.1.1 Identifying standard changes

Standard changes are often identified and managed by domain owners and/or the subject matter experts in areas such as:

- Network operations

- Applications development

- Infrastructure and server teams

- Security.

The change manager can ask the domain owners to create a listing of frequently implemented changes to be considered as standard changes. Domain owners often welcome the opportunity to increase the standard changes that can be approved locally without approaching the CAB.

The CAB should be actively looking for changes that could be made into standard changes; they should review every RFC bearing that in mind.

5.1.1.2 Candidates for standard changes

The following questions will help you identify good candidates to be standard changes:

- Is the change performed frequently in daily operation?
- Would errors relating to the change cause little impact and be easy to recover from?
- Are the means in place to record details of the changes (e.g. system logs)?
- Has the CAB found issues with similar change requests in the recent past?
  - Change requests that have significant issues discovered by the CAB in the review process may not be mature or stable enough to be good candidates.
- Have there been incidents or disruptions to service caused by similar changes in the recent past?
  - Changes that have caused post-implementation issues in the past are probably not good candidates, though it's possible that the process of documenting and reviewing these types of change – especially if the cause of previous issues has been identified and addressed – can make them suitable for consideration.
- Is there a documented process for the requested change, and is it consistently followed? (If not, could one be created?)
  - Changes that are consistently carried out following a well-understood process (even if not documented) are very promising candidates. If the process is consistently followed, documentation can be created, validated and submitted as a standard change candidate.
- Is the change associated with delivery of services listed in the service catalogue?
- Is the request associated with any critical business functions or a highly sensitive infrastructure?
  - Changes to key infrastructure or associated with the delivery of critical business services require a higher level of review and don't generally make good standard changes. This is not a hard and fast rule, however. Some organizations classify configuration items (CIs) according to their business criticality; such labelling can be used to determine whether standard changes are appropriate.
- Is the change associated with a highly automated infrastructure?

- Infrastructure components with a high degree of virtualization and automation, such as automated provision of standardized infrastructure components or automated testing and deployment of code (following DevOps or similar continuous flow methodologies), are good candidates for standard changes.

Examples include:

- PC hardware replacement/repair

- Backup and recovery

- User account management

- Daily job scheduling, processing and operational work

- Scheduled periodic system patching of non-critical systems such as PCs

- Infrastructure capacity upgrades

- Adding switch port capacity in office areas

- Most movements/additions/changes to telecoms equipment.

5.1.1.3 Making a standard change

If a change looks like a good candidate for a standard change, it should be assigned an owner who will:

- Ensure there's good documentation for how the standard change is/will be carried out.

    Documentation for standard changes varies widely from company to company. The format is largely unimportant but should specifically identify:

    - The steps required to complete the change

    - Who will be responsible for performing the process steps

    - Where and how the change status is logged and communicated to other stakeholders

    - How the change will be tested to ensure it works as planned and doesn't cause unintended consequences

    - The steps to undo the change (roll-back) if issues are discovered

- Submit the standard change documentation to the CAB as an RFC for review.

The CAB will review the proposed standard change process for risk and operational concerns, much as they would any other change request. The only difference is that they are reviewing a *process* for performing certain defined changes. The CAB must review it for risk, value realization and unintended consequences, as well as ensuring that such changes are captured for governance and compliance (often in system logs).

The appropriate change authority then approves (or rejects) the proposal for a standard change.

Once a standard change has been approved, it is carried out as needed in daily operation, following the approved process.

5.1.1.4 Operational considerations

Despite change management delegating authority to allow standard changes to be executed without hesitation as needed in daily operations, such changes remain under their authority. Nevertheless, once the risks and operational controls of a standard change have been reviewed and approved, they do not need to be revisited unless issues arise.

If any incidents do arise from approved standard changes, change management should act to determine the underlying reason(s) and appropriate corrective action.

In daily operation, standard changes should not be:

- Submitted as RFCs
- Reviewed individually by the CAB

but should be:

- Completed following the approved, documented process
- Recorded in a way that can be reviewed by operational staff (for incident and problem management, continual improvement and compliance)
- Periodically reviewed for improvements.

5.1.1.5 Standard change workgroups

Some organizations combine the elements of standard changes with delegated authority (section 5.1.2) by establishing (change management) workgroups who are tasked with managing standard changes within a given domain (e.g. network, database etc.). This is an excellent practice, especially for organizations with a large volume of changes.

Standard change workgroups are empowered to operate with coordinated independence to recommend, review, establish, oversee, revise and deny standard changes under their authority. Representatives of the various workgroups should be standing members of the CAB to facilitate two-way communication and coordination of standard changes.

The process and practice is as above, except that the tasks are carried out by the workgroup, not at the normal CAB meetings. The CAB is made aware of standard changes that are under consideration and that have been approved, but beyond that, the work is completely delegated to the workgroup.

5.1.1.6 In practice

Standard changes are an excellent way to allow the CAB to focus on much larger and more critical changes while maintaining appropriate control. Once the level of risk has been identified and accepted in the form of a standard change, change management should be keen to give operations staff the authority to use it as needed, rather than begrudgingly, as if doing them a favour.

The mature view of standard changes is that it's achieving one of the key goals of change management – appropriately managing risk (Chapter 1). Therefore, unless or until issues arise, the CAB should pay very little attention to standard changes.

If you are just getting started with formal change management, or if you have an unpopular and bureaucratic CAB, standard changes will be a welcome relief. I strongly encourage setting a goal for number or percentage of standard changes, and track that as a change management operational metric.

5.1.2 Delegated change authority

It is common for change management to directly review all changes through the CAB. In many cases, however, those involved with the development of changes are better able to effectively manage change-related risks. Delegated change authority formally recognizes and embraces this reality while maintaining effective change oversight.

As change management matures, it becomes increasingly concerned with the overall process. To focus on the bigger picture and higher-risk changes, the CAB should delegate parts of its responsibility. Establishing a clear and well-understood structure for delegating change authority is an effective way to achieve the need to manage risk without a CAB review.

Care must be taken to eliminate ambiguity and confusion with delegated authority. Whatever structure is chosen, it must be kept as simple as possible and easy to communicate. Roles and responsibilities must be clear and well understood by all staff.

As much as possible, delegated domains should be logical groupings, with few if any exceptions – for example, all network changes at a remote office could easily be delegated to the local operations manager. (Exceptions to the delegated authority structure may include response to a defined risk or incident occurrence.) Delegation must also consider the organizational culture around ownership and accountability. A typical approach is that higher levels of organizational authority are required to approve changes that have a higher risk and potential business impact.

Whatever form of change delegation is employed, it should be documented in a written change policy and communicated to all staff (see Appendix A). The policy should clearly define domains and the criteria for determining ownership of them. It should also state when other domains must be engaged or the change escalated to a higher authority. Lack of clarity in this regard is a common source of tension and ineffectiveness.

5.1.2.1 Hierarchical change delegation

The structure shown in Table 5.1 illustrates how changes with higher risk and impact must be approved at higher authority levels. It is important to note that those with higher authority don't necessarily do the detailed change and risk analysis, but because of their level in the organization (and also higher degree of accountability), they are in a position to make the important business decisions and accept responsibility for the potential of negative impact in the form of financial or other losses.

Use operational metrics (discussed in sections 3.3 and 4.4) to determine whether adjustments to delegation levels are required to maintain appropriate risk management. In concept, changes should be pushed down to as low a level in the appropriate hierarchy as possible. If, over time, certain change types have a good track record for successful implementation, change management should consider moving authority for these changes to a lower level with ongoing monitoring for continued success.

Where there's a hierarchical structure, there must be clarity regarding why and how changes are escalated from one authority to another.

Table 5.1 Hierarchical change authority

Authority level	Extent of risk/impact on the organization
Business management	High risk to entire enterprise
	Significant cost
Senior IT	Multiple IT services impacted
	Multiple lines of business impacted
Change manager	Single IT service impacted
	Single business unit impacted
Local delegated authority	Low-risk changes
	Standard changes

Another approach is delegation by domain. In Figure 5.1 you can see how small changes in each domain can be approved by an identified local authority such as the operations manager for that domain.

For changes that cross organizational boundaries – infrastructure or applications that support multiple business units – responsibility is assigned to a level that has the organizational authority to accept risk on behalf of all involved.

This model acknowledges that many changes have localized risk and can be effectively reviewed and authorized by an authority at the site.

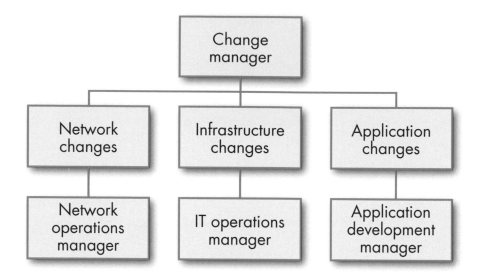

Figure 5.1 Domain delegation

For instance, a company with offices in multiple locations may have network managers at each site who have delegated change authority for network changes at that location, provided the changes do not impact on other sites. An example of this regional model can be seen in Figure 5.2.

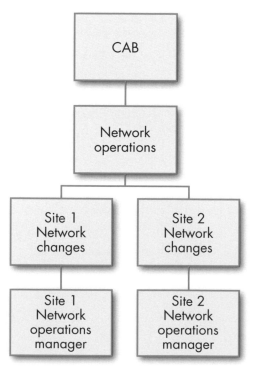

Figure 5.2 Sample regional delegation within domain

Authority for changes made to routers that connect to the wide area network (WAN) between sites must be referred upwards to an individual with the authority for all impacted locations (in some organizations this will mean the person responsible for the entire WAN).

In the same way, if systems and resources from other sites regularly access systems at the local site, authority for changes that could potentially compromise access to the local networks or systems should also reside with a higher authority.

The organization's change policy should include a provision that clearly states all delegated change authority is provisional. Each delegate must understand that if they approve changes that cause issues or negatively impact on the business, they are accountable to the delegating authority and can have their authority revoked. The tension between desire for the autonomy to move quickly to meet business need and the potential to lose that autonomy after significant failure creates an effective method of managing delegated change authority. In practice, if the delegate is not completely comfortable with a proposed change, or is not 100% convinced that there will be no impact on other parts of the business, they should escalate the decision to a higher level.

The entire change delegation must be kept as simple as possible. Well-understood domains (such as network and infrastructure) and clear criteria are key. The main requirement is that the policy be clear, understood and acknowledged by the staff.

5.1.3 Change models

Change models provide a defined approach to specific types of change. Once approved, requesters simply plug in the particulars of their situation, components and parameters.

They are a useful way of streamlining changes that are common but don't meet the criteria for standard change (usually because they're either not performed frequently or they involve higher risk). Change models can be considered for changes such as:

- Application deployments, especially where automated
- Deploying monthly PC patches
- Service requests that are not routine, such as:
  - Access requests to critical infrastructure components
  - Firewall policy changes
- Regular system maintenance such as:
  - Balancing server loads
  - Storage capacity upgrades
  - Installing new wireless access points.

In practice, change models are a generic, defined procedure that's been approved by change management for certain types of change. Once approved, change models become straightforward templates that ensure the RFC is completed with the requisite information.

The model should include:

● Process steps for completing the change

● Sequence for completing the steps, especially noting dependencies

● Generic roles and responsibilities, including:

  ● Who will perform which parts of the change

  ● What change authority approves the change

● Timeframe for change implementation, including roll-back checkpoints

● Escalation process in the event that things do not go as planned.

Change model: server decommission

An organization had several high-impact incidents that resulted from decommissioning servers that were thought to no longer be in use. Because staff were convinced that the servers were not being used, the servers were not only shut down, but physically removed. When it was discovered that the servers were still in use by production systems, it took significant time and effort to reinstall them and bring them back online.

Change management requested the server subject matter expert to propose a standard decommissioning process that would avoid a recurrence of this problem.

The process included steps to verify remote connections and activity. Only after verification that there was no remote network activity and all remote connections were severed, without impact, could the servers to be decommissioned be logically removed.

The server would be left in the standby mode for a week. After a week with no production problems, the server would be shut down.

The server would be left in this shutdown state for another week. After a second week with no problems, the server would be physically decommissioned and removed.

The steps were documented in a change model that was strictly followed. RFCs for server decommissions were filled out with the specific configuration items using the model.

Decommissioning servers went from a lengthy CAB discussion to a delegated change using a predefined model, while greatly reducing the number of related issues.

5.2 Change windows

All services require periodic maintenance and upgrading. Change windows should be established early in the design phase of a service or of critical infrastructure components. It is important for these to be negotiated early, because once a service is live, users develop expectations and find downtime for any reason to be problematic. If users know from the outset that the service will be unavailable every Sunday from 10:00 pm to 12:00 midnight (for instance), they will consider it an integral part of the service and build expectations around the window.

Change windows should be reflected in the service availability parameters included in applicable service level agreements.

Be careful to ensure there is no unexpected impact from maintenance windows. For instance, components of one service that's undergoing maintenance may be used by other services, with unintended consequences for the second service.

Establish various types of change window to meet the needs of the organization and the growing change management programme.

Weekly windows could be established for:

● Network

● Infrastructure

● Maintenance.

Specific change windows should be established for major and minor release implementations. Figure 5.3 shows what a change window calendar may look like. It shows weekly windows that are focused on a specific infrastructure component, with scheduled releases on the first weekend of the month and minor releases on the third weekend. This is by no means a recommendation but represents a typical calendar. Obviously, the needs of your organization should determine the optimal solution.

In conjunction with release management, the establishment of a release cycle is important to handle normal changes to services. A typical release schedule has a monthly release for normal service updates. Normal changes are proposed with a target release window in mind, and can be scheduled well in advance, as a normal part of release planning.

Typically, approved changes are implemented in the next scheduled change window – generally the next available after-hours slot.

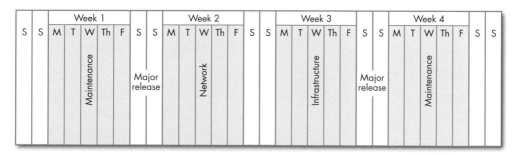

Figure 5.3 Sample change windows calendar

Emergency changes are unique in that they have high impact and urgency necessitating immediate implementation. Emergency changes can target an existing change window, or, based on business impact, can be implemented as needed with appropriate communication and coordination. Because the need for an emergency change is typically unplanned, the organization may elect to use a workaround as a temporary approach to address scheduling or other immediate issues.

5.2.1 Maintenance

Established change windows help change management understand customer requirements and expectations without having to inform customers each time a change is planned. Because customers already know the service will be down, no further customer coordination is required (as long as changes are successful within the pre-established window).

5.2.2 Change implementation staff

The staff responsible for performing the implementation of changes must be included in the process of establishing change windows. Change windows frequently require after-hours work and must be coordinated with the teams performing the work.

Not including staff in the negotiation process nearly guarantees they will feel undue pressure and resentment when asked to implement changes in timeframes they had no part in deciding. Again, establishing windows as early as possible during service design goes a long way towards gaining full support for your change programme.

5.2.3 Minimizing customer impact

Change windows are always negotiated with customers and users, and based on minimizing business impact, while maximizing the ability to effectively implement required changes. Understanding patterns of business activity is crucial to identifying change windows. All customers must be considered, as not all users have the same usage patterns.

Before establishing change windows, be sure to have them thoroughly reviewed by your technical experts who are familiar with the inner workings of the infrastructure. Also be on the lookout for maintenance windows that are mutually exclusive; for instance, a network maintenance window overlapping a server maintenance window, making the server inaccessible to its administrators.

5.2.4 Window length

The length of each change window must match the timeframe required for both the implementation and roll-back of anticipated changes. In other words, the window must be long enough for changes to be successfully completed within the given timeframe, but also for changes that are not successful to be fully rolled back before the window ends. It is also wise to build in enough buffer to allow for unexpected circumstances.

Patterns of business activity and change windows

When negotiating a weekly change window on the company's order processing system, the order processing staff were consulted, and agreed that any night after 5:30 pm would be fine. Their key concern was making sure the tool was available early in the morning so they could generate the sales reports required for daily management reporting.

It was agreed that Thursday nights at 6:00 pm would be the change window.

After several months, word got around that the field sales staff were very frustrated with the order processing system. Looking into it, it was discovered that, unlike main office staff, field sales representatives meet with customers all day, generally taking a client to dinner. After dinner, they would return to their hotel room and update their sales.

Unbeknownst to the office and IT staff, Thursday night was the cut-off for weekly sales submissions used to calculate commissions.

After talking with the sales staff and management, it was determined that it would be very rare for any sales staff to be on the road Friday night. This made Friday at 6:00 pm a much better choice for the maintenance window.

Key lesson: it is important to engage all stakeholders when negotiating change windows.

Each organization must set its own policies around change windows usage, but they must be documented and well understood by the operations staff.

5.2.5 Change windows and change planning

Having established, customer-agreed maintenance windows greatly simplifies the scheduling aspect of change evaluation. Medium- and lower-priority changes should always be performed in the agreed maintenance window.

This simplifies the RFC somewhat, as the requester simply selects which change window the proposed change is targeting for implementation. The CAB can then take a holistic view of all the proposed changes for a given window. Are there too many changes for the window? Are any mutually exclusive? If change windows are becoming overloaded, or there are conflicts, the CAB must reconcile them, based on priority and business need.

Only emergency changes should be considered for implementation outside the change window.

5.3 Chapter 5: key concepts

The key concepts in Chapter 5 can be summarized as follows:

- Reduce the number of changes coming to the CAB by:
  - Delegating change authority
  - Introducing standard changes
  - Using change models
- Establish change windows to reduce logistical coordination.

6 Adopt and adapt

No house should ever be on a hill or on anything. It should be of the hill. Belonging to it. Hill and house should live together each the happier for the other. Frank Lloyd Wright

6.1 The practitioner's dilemma

Experience tells us that a successful change management programme is based upon the practitioner's ability to successfully adopt and adapt the principles of change management to the needs of their organization. 'Adopt and adapt' has rightfully become a guiding principle in best practice. I'll make one brief note on the words: adoption is a choice that an organization makes to accept a practice, to adopt it as its own. Adaptation, on the other hand, is the process of adjusting (adapting) the adopted practice to make it work in the specific organization.

Unfortunately, there is no magic formula for doing this; no shortcut to a successful change management programme. As tempting as it may be to try implementing change management processes 'by the book', this is where many well-intended efforts go astray. Trying to 'implement' change management from best practices without considering the unique challenges of your company, the business it's in and the organizational culture is ill-advised and almost certain to deliver far less value than anticipated.

This is the challenge faced by the IT service management (ITSM) practitioner – how to make change management work in the 'real world' – in your company.

The successful practitioner must have:

● A clear understanding of the organizational objectives for change management

● The knowledge, skills and experience in ITSM best practices and standards

● Skills in adapting ITSM to fit the needs of the organization

● Skills in managing organizational change.

The practitioner must expertly synthesize these in the right combination in order to put together a successful change management programme.

6.1.1 Understanding the organization's change management objectives

It stands to reason: if you don't know what you're trying to accomplish, you're not very likely to achieve it. The only legitimate reason to embark on a change management programme is to achieve company objectives. What are those objectives, in specific, measurable terms?

As I mentioned previously, change management efforts are often started on the heels of one or more significant or impactful outages believed to have been caused by poorly managed changes. And while effective change management can help organizations in many ways, it's not a panacea.

Care must be taken to ensure the organization's expectations of the change management programme are realistic and do not exceed the limits of what change management, as a practice, can offer. This point is often given minimal attention, if any, in the rush to get on with implementing a change programme. I encourage you to carefully consider what the organization hopes to achieve with change management and ensure the outcomes are realistic and attainable with the proposed effort.

Recall from Chapter 1 that change management seeks to:

- Support timely and effective implementation of business-required changes

- Appropriately manage risk to the business

- Minimize negative impact of changes to/for the business

- Ensure changes achieve desired business outcomes

- Ensure governance and compliance expectations are met.

Keep in mind that all these objectives tie directly to managing changes. In other words, change management does not seek to manage all IT risk and impact – only change-related risks and negative impact.

6.1.2 ITSM knowledge and experience

Knowledge, skills and experience in ITSM frameworks and standards are essential for the ITSM practitioner to be successful in implementing an effective change management programme. You should have training and practical experience in both best-practice frameworks such as ITIL and COBIT and standards such as ISO 20000.

I encourage you to become deeply familiar with the fundamentals of change management from multiple perspectives. You would do well to pursue foundation and advanced training and certification in the related frameworks and standards. Develop professional relationships with other change managers and those with practical experience in adopting change management.

Understand more; implement less

In the early stage of planning your change management programme, learn everything you can about the transition phase in general, and change management in particular. Seek to understand the big picture, paying special attention to understand the outcomes and business value of change management.

Don't assume you know how it should work. You may have experience with change management in another organization, but every organization is unique.

Immerse yourself in all things change management. Learn from those who are doing it, and don't assume that they are fanatical because they are talking about and doing things that you don't yet understand.

If you understand just enough when you begin with change management, you are more likely to do things that you'll later regret, or that make future iterations harder.

Think about how change management will fit with other processes that may be implemented later – configuration management, release management. Think too how DevOps and Agile will influence it.

Once you have a solid understanding, implement far less than you now understand. Your deep knowledge is what will help you be successful in the initial stage. While it's very tempting to do more, it's far better to be successful with a small programme and make incremental changes going forward.

It is also important to understand the difference between standards and best-practice frameworks.

Frameworks are a collection of best practices structured such that they suggest what needs to happen to achieve the objectives of each capability.

Standards, by contrast, describe what elements must be in place to successfully achieve the capability. Whereas frameworks are suggestive, standards are designed to be benchmarked against. For example, organizations can pursue ISO 20000 certification as a means of demonstrating compliance with the ISO ITSM standard.

The practitioner should be familiar with both frameworks and standards, especially as they adapt processes.

6.1.3 Adapting ITSM

All processes require adaptation to work in a given company environment. There's no such thing as a 'standard' IT organization, and there's no such thing as a 'by the book' implementation of a best-practice process.

Best practices being collected from successful organizations, it's not at all surprising to find many elements of change management being practised in most of them. Where process elements are present, they should be acknowledged and readily embraced as how the organization is fulfilling that part of the process. It is not necessary to change the name of existing practices to the ones suggested in the frameworks, either. Simply design your change programme around those existing elements that are working well.

> **You're doing everything wrong**
> An organization had commissioned a process maturity assessment that indicated a maturity level for change management of zero. There was little evidence of any formal change management, most notable of which was the absence of a change advisory board (CAB). The assessment of zero was accepted and change management implemented.
>
> There was a growing resentment and cultural resistance to the new change programme, and it took the change manager months to realize that many of the fundamentals of change management were indeed being done already, just in isolated, informal processes.
>
> IT staff felt that their existing efforts to manage changes had been disregarded and were being replaced without acknowledgement of the value they had brought to the business. In short, the message they felt they had been given was, 'You're doing it all wrong', which, of course, was not at all the intent.
>
> In time, the change manager recognized that many parts of change management were already in place, and moved to embrace them, but staff morale had already been undermined.
>
> A more comprehensive inventory of existing practices would have gone a long way to identify and allow the incorporation of existing practices that were working.

A word of caution on embracing existing processes: I've seen too many organizations attempting to use ITSM terms in ways that are significantly out of sync with the spirit of change management. Changing the name of existing practices does not mean they are fulfilling the intended purpose. Using ITSM terms to describe process elements that don't fulfil the objectives for that process also gives a false sense that you're 'doing change management', when in fact, you may be falling far short of its requirements.

This is why the focus of change management must be on the end-to-end lifecycle of changes. It starts with an identified need for a change. Once the need has been documented in an RFC, it enters into the change lifecycle. Change management is responsible for managing the change through the process from idea to fully realized business value from the implemented change. Along the way, change management must facilitate some key decisions on risk, business value and operational considerations. The end result of the change management capability is business-required outcomes.

Just because you have a CAB does not mean it is fulfilling the objectives of change management. (By the same token, outcomes can be achieved without a CAB, or other process elements.) The point here is that you shouldn't adopt new (ITSM) terminology for the same old practices if those practices aren't achieving the desired business outcomes.

Having said that, however, there are limits to how processes should be adapted. Process adaptation is neither a licence to do violence to the spirit and intent of the process nor to mangle it beyond recognition.

Let's look at some poor practices in process adaptation.

While the practitioner must modify the process in ways that make it work in the organization, adapting a process does not mean simply throwing out parts that you don't understand or care for.

Best practices are distilled down to the most essential elements commonly practised in successful organizations. Each element is included for specific reasons, not all of which are intuitively obvious, especially in the early stages of adoption. In many cases, those parts that don't seem to make sense are associated with other processes that have not yet been implemented (and may never be, depending on the organization's needs).

Before eliminating parts of a process, carefully consider:

- Does it support any of the organization's stated goals for change management?

- Is it already being performed in the organization (and can it be embraced)?

- Can it be modified to work better in the organization (rather than eliminated)?

- Does it support a compliance objective (governance, regulatory or audit)?

This is where knowledge of standards such as ISO 20000 or more prescriptive frameworks such as COBIT can be helpful. In these, you will find core elements that must be present for it to be considered that changes are being managed effectively. Think very hard before eliminating any of these core functions.

In some cases, it may be better to implement the part in question at a very basic level rather than eliminate it altogether. Building in a 'placeholder' for future enhancements can make it easier to mature the process when the time comes.

6.1.4 Organizational change management

Organizational change is an area of study all to itself, but it is so critical to successfully adopting a change programme that I'll mention it briefly here.

As I said before, organizational change management deals with the people side of changes – how to successfully transition people and organizations from one state to another. In *Managing Transitions*, William Bridges (2009) describes three major phases of personal and organizational transition:

- Ending – letting go

- Neutral zone

- New beginnings – seeing new possibilities.

Bridges suggests: to help people move from the old state to the new, it is important to transition through these three phases. All too often, when implementing IT change management, the focus is primarily on the process aspects at the expense of the people. This is a major source of failure when implementing a change programme.

It is important to acknowledge what is ending as you're implementing a new change management programme. In some cases there's loss of autonomy in managing changes. Staff may have fond memories of times they successfully implemented changes through heroics and camaraderie that they feel is now being replaced with a cold process. In many organizations, senior staff are trusted to implement beneficial changes and allowed to implement and fix changes with little or no oversight.

Whether these are real or perceived losses, the impact on people is the same, and wise is the change practitioner who is mindful of, and acknowledges, these losses. It is important to recognize the work staff have done to implement changes and protect the customer from inconvenience, or worse.

This can be tricky if your change programme is precipitated (as it often is) by some rather embarrassing change-related outages, and management is demanding better change management. The messaging here is very important. Stories are part of organizational culture. It's helpful to let staff recollect and celebrate past change efforts – perhaps particularly difficult implementations, or times when they delivered changes under extreme pressure or to unreasonable timescales – and came through for the business.

If you start your change programme with 'your change management is bad, and management is demanding we fix it', be prepared for a lengthy battle in implementing change management. The process may be implemented but, as Bridges points out, people will not have transitioned.

The neutral zone is entered when people begin to realize that change is happening. In this phase people can feel demotivated and anxious. This is where a lot of resistance begins – as people resent the change and try to protect what was.

To help people move forward through the neutral zone, they need to understand why the change is necessary, how it fits into the big picture. Here again, it's important to focus on the business outcomes, not the process. The business outcomes of change management are hard to disagree with. Talk about these outcomes more than the process details. It is also helpful to emphasize what is in it for them.

The last phase of Bridges' organizational change management model is where people start seeing new possibilities in the change. As you can imagine, when staff enter this phase they become powerful allies to help achieve the objectives of change management. This is where you want people to arrive. I've included a lot of people-related aspects of implementing a change programme throughout this publication for this very reason.

6.2 Understanding best practices

Some will argue that best practices, because they're aggregated from a large set of organizations, are, by definition, average. Adopting them, so the theory goes, can only make you average. That point of view misses the point. Use best practices as the starting point to provide the most effective and efficient way to produce a particular outcome. This revised view makes the application of best practice hardly average – it supports maturing the organization to be more competitive.

Best practices are sometimes seen as 'inside the box' thinking, where 'the box' has become the standard against which creative thinking is measured. This type of thinking is basically solving problems from within the bounds of a well-understood set of solutions for a given field.

Some even go as far as equating best practices with copying others, a view that will not motivate innovative staff to embrace them. While this argument may seem logical on the surface, it doesn't hold up under closer examination.

With an unapologetic nod to Jim Collins' *Good to Great* (2001), I maintain that establishing a solid 'good' foundation with best practices frees up resources to solve your unique problems. Great companies don't try to be great at everything; they focus their resources on what's most important. That being the case, why would you spend more than minimal resources solving problems the industry has already solved? Just adopt and go.

The concept is everywhere, hidden in plain sight:

- Does a solid training in musical notation and harmony theory limit a musician to being average?

- Do building codes limit architects to the design of uninspiring, average buildings?

- Do creative civil engineers ignore the guidance in the *Standard Handbook for Civil Engineers* when designing bridges?

- Do professional athletes and their trainers ignore current practices in training and nutrition?

The answer is obvious: they don't, and that's exactly the point. Everyone in these fields has access to the same standard best-practice materials. They give guidance on how things are done – a foundation of sorts.

Reinventing the wheel

Did you know that there are thousands of patents for wheels? You could be forgiven for thinking that the wheel was being reinvented. But it's really not. In practice, the wheel concept is being adapted in creative ways to solve new and unique challenges.

Imagine the effort it would require to invent a completely new transport method without adapting the wheel in some way. I'm quite sure it's possible, but why would you do that when you could simply take the wheel, adapt it as needed, and have a quick solution that's based on well-tested principles?

Excellence outside the box

Every organization has limited resources, and each must decide how to use the resources they have to best achieve their goals.

Leading organizations are very intentional about where they spend their resources. They know full well that you won't lead markets or be excellent by mimicking others. This is exactly why they don't waste time and effort on problems that have already been solved, but instead leverage best practices to quickly build a solid 'good' platform. In so doing, they free up their talent and other resources to work on those challenges that are unique to their business which will give them the advantage they need to compete in the marketplace and make their enterprise truly great.

Far from limiting you to being average, best practices, when used correctly, are the fastest way to excellence.

But what makes some great and others average is what they do above the foundation of these best practices. To be great, you have to build on best practices, not ignore them.

I contend that best practices do not make an organization great; they reduce the effort required to be good. They are the proverbial wheel that needn't be reinvented.

6.2.1 Adapting change management

Understanding applicable standards and the associated best practices is invaluable to the practitioner in achieving change management success. However, there are some poor practices that go under the umbrella of 'adapt' that I want to address specifically.

6.2.1.1 What adapting is not

Adapting does not give licence to eliminate the core functionality of change management without considering how its removal will impact on the outcomes. This is one of the key reasons I've consistently emphasized the outcomes over the process. If elements of the process need to be modified or removed and the objectives are still met, then you're in a safe space.

The practitioner must know what elements can be modified and/or eliminated and still deliver the required outcomes, or else the ability to mature when required could be unduly hampered. Excluding certain elements will jeopardize the realization of desired outcomes.

Adapting is not importing new terminology to use with the same old practices. Here again is why I focus on outcomes. It doesn't really matter what you call elements of change management – if they're not producing the needed outcomes, then nothing you call them can make them right. And by the same token, if they are producing the desired outcomes, the 'wrong' names are irrelevant as well.

Adapting is also not making an organization follow the practices of some other organization you may have worked in (as an employee or consultant). Familiarity with a successful change programme elsewhere does not mean it's the proper approach in the current company. Adaptation must be done intentionally to make change management work in the environment and culture of the organization adapting it. It must address the specific circumstances of the target environment.

6.2.2 Continual service improvement

One of the key strategies for successful adoption of change management has roots in continual service improvement (CSI). CSI, in short, is an approach to incremental improvement. It's especially powerful for adopting and adapting a new process because it starts with a minimal process and provides both staff involvement and process evolution – which does wonders for acceptance of change within the organization.

This method of adoption is based on Kotter's model for achieving change (Kotter, 1996). Obviously, I can't do it justice in a short section here, but the key steps are:

- Build a sense of urgency

- Form a coalition

- Create a vision for change

- Communicate the change

- Empower action

- Create 'quick wins'

- Build on the change

- Keep the momentum.

The first three are investing in creating an environment that supports change, while the next three encourages engagement with the change. The final two encourage sustaining the momentum gained (and not drifting back to the original state.)

As I've said time and again – if it's a change process you want, you can have that in a few days. But if it's a change capability (including the cultural changes required for success), that's going to take a bit more time. More time, because culture takes time to change, and process changes are deeply rooted in organizational culture.

The hardest part of adopting change management (or any other process for that matter) is changing the attitudes and behaviours that make up the culture.

CSI is an excellent way of getting the people who can either make or break your change programme on board. You'll know you're making progress when staff propose changes to the change programme that make it more effective or efficient.

6.3 Chapter 6: key concepts

The key concepts in Chapter 6 can be summarized as follows:

- Change management must be adapted to fit the organization.

- Multiple frameworks and standards should be consulted.

- Be cautious about eliminating parts of best practices; consider limited implementation rather than complete elimination.

- Leverage continual service improvement methodology as an adoption and buy-in strategy.

7 Future of change management

Isn't it funny how day by day nothing changes, but when you look back, everything is different … C. S. Lewis

Change management has existed in some form since the very early days of IT, when systems were monolithic and changes were relatively infrequent but well planned. The evolution in change management since then is about doing the same thing faster, within the complexities of modern computing.

With the advent of radically new approaches in development methodologies, change can no longer be the same thing only different – change management must itself change to meet the needs of the organization. The goals for change management remain: they must always be clear and focused on business outcomes.

The goals, again, are to:

● Support timely and effective implementation of business-required changes

● Appropriately manage risk to the business

● Minimize the negative impact of changes to/for the business

● Ensure changes achieve the desired business outcomes

● Ensure governance and compliance expectations are met.

In Chapters 3, 4 and 5, I've outlined a practical approach to building an effective change management capability that has a high likelihood of being successful in your organization. I've leveraged best practices in ITSM and a proven, common-sense approach to meet these core goals.

But if we take a closer look at the goals, you'll see that some things have not been addressed. Many things, in fact, such as weekly change advisory board (CAB) meetings, change schedules and lots of things relating to processes. I've emphasized the need to focus on business outcomes and not process, and I did so for a reason.

In ITSM, we use the term 'fit for purpose' as a way of describing if something is fully suitable for the purpose for which it was designed. We must continually evaluate whether our change capability is the optimal way to meet the specified goals.

No matter where your organization is with change management, you will need to continue to improve and mature it. I've shown a multiphased approach in Chapters 3 to 5, and by following it you'll have a solid change capability. You'll also have a foundation upon which you can continue to grow with the business.

7.1 A brief introduction to the new methodologies

Traditional application development followed a waterfall methodology, where development cycles could be lengthy, and the objective was to deliver a completely functional application that met the documented requirements. It was called waterfall because the process flowed in a linear step-wise process from requirements and design through development and testing and then to final release. It borrowed heavily from engineering methodologies used to design buildings and bridges. The quality of the result followed directly from the discipline and precision applied to each step, with subsequent steps being built on the solid foundation of the previous step.

This methodology fitted very well with what I will call traditional change management. The relatively brief delay to bring changes to the CAB for review was minor in comparison with their lengthy development cycles. Many corporations still rely on applications developed using this methodology decades ago, but the challenges and rate of change required to support the business has increased so dramatically of late that, quite frankly, change management by and large hasn't kept pace.

Enter the variety of new development methodologies, most of which fall under the generalized description of 'iterative development'. While each has its own unique focus, they all share some common objectives:

- More frequent releases of usable code
- Shorter development cycles focusing on the delivery of incremental and agreed functionality
- Higher degree of business engagement
- Higher use of automation (especially building and testing).

Of the many iterative approaches, the ones that are getting the bulk of attention are Agile and DevOps. Keep in mind that these approaches are rapidly evolving, so as soon as this guidance is published, the information will quickly date. However, the implications for change management will remain the same.

7.1.1 Agile

I won't attempt to do justice to the Agile approach to development. I'm sure those familiar with its use will take issue with my brief summary. However, as the point is to highlight its implication for change management – especially as it has traditionally been practised – I think it serves my purpose. (For a more complete treatment, please check out the Agile Manifesto and related information; see Bibliography.)

Agile is a development approach that seeks to release small incremental units of functional code rapidly. Each release is functional and fully tested, adding incremental and usable functionality to the system (or application) under development. The development cycle varies depending on the organization, but is typically 1–4 weeks. While each increment should be stand-alone, it is likely that multiple increments of functionality will be combined into a single release package. The release rate is aided by using automated testing and build automation.

To summarize, Agile:

- Is a development approach

- Involves small, frequent releases (i.e. iterations or sprints)

- Requires as a minimum the highest-priority functionality to meet customer needs first (minimal viable product)

- Is delivered by a cross-functional team in a short period of time, typically 1–4 weeks (note that this means functionality is delivered is small units).

7.1.2 DevOps

DevOps is similar to Agile in that it features short iterations of development activity, the result of which is the delivery of a usable subset of the required features. It incorporates as a core value the concepts of Lean IT in its relentless pursuit of rapid delivery of business value. DevOps practitioners acknowledge that it is a movement that requires organizational culture to be addressed.

The DevOps approach centres around three core 'ways' of adoption:

- Systems thinking (emphasizing the performance of the whole system)

- Amplifying all feedback loops (to improve understanding)

- Embedding a culture of continual experimentation and learning.

It emphasizes the highly collaborative involvement of both applications development and operations staff throughout the development process.

In practice, DevOps relies heavily on automation – in the building and testing processes and self-service infrastructure on demand – as well as during promotion to production. DevOps seeks to create a collaborative culture in which IT and business resources work together to maximize rapid delivery of business value.

A related concept is continual delivery – the idea that any given change can be released as a stable, usable application, and there can be many such releases daily. In this regard, DevOps borrows heavily from *Theory of Constraints* (Goldratt and Cox, 1990) and views the entire development lifecycle as a continuous flow, where work should flow smoothly 'from left to right' (i.e. from idea to business value).

Following these concepts and methods, leading organizations are able to release potentially hundreds of changes per day. Obviously, traditional (CAB-based) change management is ill-equipped to function at this pace, let alone add much value.

7.2 What's changed; what's the same

What has changed is the challenges to change management that are associated with iterative methodologies, such as:

● Supporting frequent, small releases

● Integrating automated testing into the lifecycle

● Blurring the lines between business and IT.

What change management and the new approaches have in common is that they:

● Support timely and effective implementation of business-required changes

● Minimize any negative impact and manage risk

● Ensure realization of business outcomes by testing changes and releases before implementation

● Ensure enterprise governance and compliance requirements are met.

7.3 Change management challenges

If change management is supplanted or optimized to DevOps development, how does it address non-software changes such as network, infrastructure, servers, firewalls and the like?

Can non-software changes be developed with DevOps? While I'm sure they can benefit from the Lean thinking and laser-like focus on business value (as I've already emphasized), there is no single approach to managing the myriad changes the modern IT shop faces.

Many organizations struggle with effective configuration management, upon which modern change management strongly depends. IT organizations hoping to bring a DevOps mindset to all changes must invest heavily in infrastructure tools to vastly reduce the effort required to manage configurations. Having effective automated tools to manage configuration opens the door to change modelling and automated testing of proposed changes.

Cloud-based services with vastly reduced organizational focus on the infrastructure also open up new opportunities for accelerating time to value of changes.

Remember W. Edwards Deming's view on engineering quality into the process rather than inspecting it in at the end. As change management matures, it must loosen its grip on individual RFCs and increase its focus on engineering (change) quality into the overall development cycle. This is where DevOps and mature change management, as I've described it in this publication, are in complete alignment.

Change authority must be pushed down as close to the development process as possible. Where there's effective (DevOps) development, build and testing, automated to the point where quality can be predicted mathematically, change authority should be delegated, essentially to the DevOps process itself. Recall that the goals of change management never required formal/human review of changes, only that change management is tasked with ensuring those objectives are being achieved. When properly designed and optimized, the development process itself ensures the goals of change management are fulfilled.

With that in mind, change management must look at the overall development capability of the organization, assessing its ability to produce the level of change quality required by the business, and make investments to bring about improvements where needed.

Change management must be constantly evaluating the results of the change capability (realization of business outcomes), and, just like our colleagues in manufacturing engineering, seek to identify where in the development process defects could most effectively be engineered out.

7.4 An effective change capability

The approach presented in this publication is designed to bring about mature change management. There's a natural progression from introducing the concept of change control, to building a mature change capability. Chapter 5 includes key elements for optimization that set the stage for a step in evolution, namely:

- Delegated change authority
- Standard changes
- Change models.

With these, change management is no longer a one-size-fits-all proposition. It now has the tools to apply the required level of oversight to the various value streams over which it has authority.

Whereas a CAB-centric change programme is problematic for high-velocity, highly automated development methodologies, these three elements are key. They allow application of the right level of autonomy to development approaches that produce quality changes in alignment with organizational needs. This is effective change management in a DevOps world. The result is a change capability that's suited to add value through effectively managing the organization's change capability.

Any organization adopting change management at this point would be ill-advised to implement a traditional, bureaucratic change process that amasses control and decision authority into a single, central body such as the CAB.

The goal of change management is an organizational capability that produces business outcomes while ensuring the principles of change management are satisfied along the way. These principles are to:

● Support timely and effective implementation of business-required changes

● Appropriately manage risk to the business

● Minimize negative impact of changes to/for the business

● Ensure changes achieve desired business outcomes

● Ensure governance and compliance expectations are met

which are served well by:

● Becoming an integral part of the IT value stream

● Letting go of the 'silo of the CAB'

● Focusing on achieving business outcomes

● Measuring itself by the business value it produces

● Achieving its goals as part of development flow.

These principles will never go out of style because they don't dictate any particular process or practice. Change management practitioners must learn to constantly adapt to meet the needs of the businesses they serve.

7.5 Where do we go from here?

Well, for starters, the need for effective management of changes isn't going to change. The challenge change management faces is its well-earned reputation of being slow, bureaucratic and hopelessly outdated. It's a legacy that requires more than minor tweaking to shed.

Change management must become a core organizational capability that never rests on prior success. What may have worked in the past won't meet the needs of the future. For most change management practitioners, the journey has only just begun. The cadence of the phased approach will continue into the future as they strive to meet their organization's ever-changing need for effective change management.

So, focus on the business, and how you can help achieve its goals. Keep up the relentless pursuit of excellence, practitioners!

Appendix A Example of a change policy

Data element	Description	Comments
Purpose	<Organization> requires an adaptive and responsive change management capability that meets the business's need for timely and effective implementation of changes to meet rapidly evolving business needs and maintain a stable and secure IT infrastructure.	
	This document describes <organization's> expectations for IT change management for the efficient and effective management of the IT changes required to maintain an IT infrastructure suitable to meet the evolving needs of <organization>.	
Scope	The scope of this policy is all changes to production IT services, regardless of request source, size, initiator or implementer, implementing or overseeing those changes.	
	Note that changes to the development and test environment are *not* subject to this policy.	
General expectations	All changes will be managed in compliance with the spirit and intent of this policy.	
	All changes must be proposed and reviewed prior to beginning development.	
	Once a change has been developed and tested, it must be reviewed and approved before release into the production environment.	
Emergency changes	Emergency changes are provided for in this policy. Emergency changes are those that are required for incidents that have significant impact on the company's finances or public perception, or can result in risk to life or bodily harm.	
	Emergency changes require the existence of a major incident, with an incident manager directing remediation efforts. The incident manager works in close coordination with the change manager to ensure emergency changes receive due oversight and are approved by the appropriate change authority.	

Table continues

109

Appendix A *continued*

Data element	Description	Comments
Standard changes	Standard changes are changes that are pre-approved, and are implemented as needed in daily operations. Standard changes are a form of delegated change authority, granted upon a CAB review and the change manager approving the proposed process for handling them. Standard changes must be implemented in a way that is materially consistent with the approved procedure. If incidents or operational issues arise from standard changes, the change manager may request a review of the process and, in conjunction with the CAB, recommend changes to the process. Standard changes can be revoked by the change manager.	
Unauthorized changes	Because of the high potential for negative impact on IT services and infrastructure stability, all changes must be approved by change management. All changes that have not been properly reviewed and approved are considered unauthorized. Unauthorized changes discovered will be treated as an incident, and are generally rolled back, unless it is determined that doing so will have undue negative impact on the business. In all cases, staff found responsible for unauthorized changes may be disciplined, as per company policy.	
Change windows	Periodic change windows shall be established in conjunction with the business for the purpose of system maintenance and change implementation. There shall be both monthly and weekly release windows (for major applications/services and for periodic maintenance/minor releases, respectively).	

Appendix B Example of a basic request for change form

Data element	Description	Comments
Tracking number	Unique ID number assigned by change manager/CAB	
Status	Current status of request. Field managed by change manager/CAB Submitted – RFC has been accepted by change manager/CAB for consideration Approved – RFC has been reviewed and approved by change manager/CAB Rejected – RFC has been reviewed and rejected by change manager/CAB – explanation provided in Comments field Pending – RFC has been reviewed and is awaiting further information, explanation or other	
Change type	Normal – Normal change Standard – Standard, pre-approved change Emergency – Changes that must happen faster than a normal CAB schedule allows. Typically associated with either rapidly emerging business needs or incident management	
Priority	Urgent – High – Medium – Low –	
Initiator of RFC	Name and department/role of the person initiating the RFC	
Submission date	Date the RFC is submitted	
Short description of change	Description of the change being requested	
Reason for change	Description of why the change is being requested, explaining the expected results and benefits from the business perspective	For significant changes, this can be a pointer to a more complete plan

Table continues

Appendix B *continued*

Data element	Description	Comments
Configuration items (CIs)	List of all CIs impacted by the change	CIs are components of the IT infrastructure (both hardware and software) that are managed by the IT organization
Customers and users impacted	Which customers and users are impacted by the proposed change. Includes those who may be potentially impacted	
Proposed change date	Desired implementation date of the proposed change	
Proposed change time	Desired time to implement the change	
Change duration	Estimated time to fully implement the change, including time to fully roll back if needed	
Approved change date	Agreed implementation date for the approved change (decided by change manager/CAB)	If organization has established release windows, substitute to match
Approved change time	Agreed implementation time for the approved change (decided by change manager/CAB)	
Implementation plan	Detailed plan to implement the proposed change Description of the major steps to be performed Who will be performing the steps	For large or high-impact changes, this may be a pointer to release or project plans. The level of detail depends on the culture as well as the nature of the requested change. Encourage requesters to err on the side of more detail
Communication plan	What information will be communicated to customers, who will receive it, and how it will be done	
Post-implementation test plan	How the change will be tested once complete to ensure the desired outcomes will be realized by the business	If organization has customer acceptance plans, include here

Data element	Description	Comments
Back-out plan	Detailed plan to undo (roll back) the change to the pre-change state in the event of trouble or unintended results	For large or high-impact changes, this may be a pointer to release or project plans
Comments	Additional information as needed	Use to document 'pending' or 'rejected' requests
Result	Status indicating the result of the change: • Success – change worked as planned • Failed – change failed, customer impacted • Rolled back – change failed, rolled back without customer impact	
Post-implementation review	Comments on the CAB's review of the change, including any lessons learned or opportunities for improvement	

Appendix C Example of a phase 2 request for change form

Data element	Description	Comments
Tracking number	Unique ID number assigned by change manager/CAB	
Description	High-level description of the proposed change in business terms	
Business case	Summary business case stating the intended value to be realized	
Business outcomes	Description of the business outcomes anticipated from the proposed change	Outcomes should be specific and measurable
Elements to be changed	Identification (at a high level) of which IT services and infrastructure components are involved with the proposed change	
Reason for change	The business case supporting the change proposal, in business terms, including the desired outcomes the business intends to achieve	
Change category	High-level category of change – maintenance, major etc.	
Risk management	Initial assessment of risks and high-level plan to manage	
Change window	The change window the change is targeting for release	
Acceptance criteria	Description of the specific test results that will define success	Should be specific and measurable. Criteria should be tied to desired outcomes (above)
Roll-back plan	High-level plan to correct or roll back the change proposal	
Implementation schedule	High-level timeline of implementation date(s)	
High-level architecture	Block diagram showing the service components, both new and existing, that are involved with the proposed change	

Bibliography

The Agile Manifesto: http://agilemanifesto.org

Bell, S. C. and Orzen, M. A. (2010). *Lean IT: Enabling and Sustaining Your Lean Transformation* (1st ed.). Productivity Press, New York.

Bridges, W. (2009). *Managing Transitions: Making the Most of Change* (3rd edn). Nicholas Brealey, London.

Collins, J. (2001). *Good to Great.* William Collins, New York.

Edwards Deming, W. (2000). *Out of the Crisis.* MIT Press, Massachusetts.

Goldratt, E. M. (1990). *Theory of Constraints.* North River Press, Massachusetts.

Goldratt, E. M. and Cox, J. (2014). *The Goal: A Process of Ongoing Improvement* (30th edn). Routledge, New York.

Kim, G., Behr, K. and Spafford, G. (2013). *The Phoenix Project: A Novel about IT, DevOps, and Helping Your Business Win.* IT Revolution Press, Portland, Oregon.

Kim, G., Humble, J., Debois, P. and Willis, J. (2016). *The DevOps Handbook: How to Create World-class Agility, Reliability, & Security in Technology Organizations.* IT Revolution Press, Portland, Oregon.

Kotter, J. P. (1996). *Leading Change.* Harvard Business Review Press, Massachusetts.

Wilkinson, P., and Schilt, J. (2008). *ABC of ICT: An Introduction.* Van Haren Publishing, Zaltbommel, Netherlands.

Index

Page numbers given in *italics* refer to tables or figures

PIR (post-implementation reviews) 48, 77
planning 91
PMO (project management offices) 65
post-implementation reviews (PIR) 48, 77
principles of change management 108
priorities of change 41–2
process flow diagrams *37, 64, 69*
process owners 42
processes
    and outcomes 10–11
    phase 1 36–53
    phase 2 64–77
project management offices (PMO) 65

quality engineering 19–20
quick wins 14

RACI (responsible, accountable, consulted, informed) *26,* 26–7
regulatory compliance 6, 60
reporting 55
requests for change (RFC) 40
    change advisory boards (CAB) 46–7
    as a decision support tool 67–8
    examples of forms 111–13, 115
    status tracking 74–7
    types *41*
responsible, accountable, consulted, informed (RACI) *26,* 26–7
reviews 36
RFC *see* requests for change
risk management 2–4
roles and responsibilities 26–7, 42, *42*

scope of change management 60–2
SDLC (systems development lifecycle) methodology 65
server decommissioning 88
simplicity (as a guiding principle) 24
skills required of a change practitioner 93–8
software updates 6
stabilization periods 76
staff for change implementation 90
stakeholder engagement 32–3
standard change workgroups 83
standard changes 48, 79–84
starting big 20–1, *21*
starting small *18,* 18–19
status tracking 74–7
success criteria 55–6
systems development lifecycle (SDLC) methodology 65

test environments 61
timeliness and change management 2
tools 24–6

unintended consequences, avoidance of 4–5
urgent changes 49–50

Waterfall methodology 104
workgroups for standard changes 83